Numerical Hamiltonian Problems

Numerical Hamiltonian Problems

J. M. Sanz-Serna
Departamento de Matemáticas
Universidad Carlos III de Madrid

M. P. Calvo
IMUVA y Departamento de Matemática Aplicada
Universidad de Valladolid

Dover Publications, Inc.
Mineola, New York

Copyright
Copyright © 1994 by J. M. Sanz-Serna and M. P. Calvo
All rights reserved.

Bibliographical Note
This Dover edition, first published in 2018, is an unabridged republication of the corrected edition published by Chapman and Hall/CRC Press, London, in 1994.

International Standard Book Number
ISBN-13: 978-0-486-82410-9
ISBN-10: 0-486-82410-1

Manufactured in the United States by LSC Communications
82410101   2018
www.doverpublications.com

Contents

Preface	xi

1	**Hamiltonian systems**	1
1.1	Hamiltonian systems	1
1.2	Examples of Hamiltonian systems	3
	1.2.1 The harmonic oscillator	3
	1.2.2 The pendulum	3
	1.2.3 The double harmonic oscillator	5
	1.2.4 Kepler's problem	6
	1.2.5 A modified Kepler problem	8
	1.2.6 Hénon-Heiles Hamiltonian	12

2	**Symplecticness**	15
2.1	The solution operator	15
2.2	Preservation of area	16
	2.2.1 Concept of preservation of area	16
	2.2.2 Preservation of area and dynamics	18
	2.2.3 Preservation of area as a characteristic property	19
2.3	Checking preservation of area: Jacobians	19
2.4	Checking preservation of area: differential forms	20
2.5	Symplectic transformations	21
2.6	Conservation of volume	23

3	**Numerical methods**	25
3.1	Numerical integrators	25
3.2	Stiff problems	27
3.3	Runge-Kutta methods	28
	3.3.1 The class of Runge-Kutta methods	28
	3.3.2 Collocation methods. Gauss methods	30

	3.3.3 Existence and uniqueness of solutions in implicit methods	33
3.4	Partitioned Runge-Kutta methods	34
3.5	Runge-Kutta-Nyström methods	36
3.6	Composition of methods. Adjoints	37
	3.6.1 Composing methods	37
	3.6.2 Adjoints	38
	3.6.3 Finding Runge-Kutta, Partitioned Runge-Kutta and Runge-Kutta-Nyström adjoints	39

4 Order conditions — 41
4.1 The order in Runge-Kutta methods — 41
4.2 The local error in Runge-Kutta methods — 43
4.3 The order in Partitioned Runge-Kutta methods — 45
4.4 The local error in Partitioned Runge-Kutta methods — 46
4.5 The order in Runge-Kutta-Nyström methods — 48
4.6 The local error in Runge-Kutta-Nyström methods — 50

5 Implementation — 53
5.1 Variable step sizes — 53
5.2 Embedded pairs — 54
5.3 Numerical experience with variable step sizes — 56
5.4 Implementing implicit methods — 61
    5.4.1 Reformulation of the equations — 61
    5.4.2 Solving the equations: functional iteration — 62
    5.4.3 Solving the equations: Newton-like iteration — 62
    5.4.4 Starting the iterations — 63
    5.4.5 Stopping the iterations — 64
    5.4.6 The algebraic equations in the Runge-Kutta-Nyström case — 64
5.5 Fourth-order Gauss method — 65

6 Symplectic integration — 69
6.1 Symplectic methods — 69
6.2 Symplectic Runge-Kutta methods — 72
6.3 Symplectic Partitioned Runge-Kutta methods — 76
6.4 Symplectic Runge-Kutta-Nyström methods — 77
6.5 Necessity of the symplecticness conditions — 80
    6.5.1 Preliminaries — 80
    6.5.2 Independence of the elementary differentials in Hamiltonian problems — 81

CONTENTS

	6.5.3 Necessity of the symplecticness conditions in the Partitioned Runge-Kutta case	82
	6.5.4 Other cases	85

7 Symplectic order conditions   **87**
7.1 Preliminaries   87
7.2 Order of symplectic Runge-Kutta methods   88
7.3 Order of symplectic Partitioned methods   91
7.4 Order of symplectic Runge-Kutta-Nyström methods   93
7.5 Homogeneous form of the order conditions   95
    7.5.1 Motivation   95
    7.5.2 The Partitioned Runge-Kutta case   95
    7.5.3 Other cases   97

8 Available symplectic methods   **99**
8.1 Symplecticness of the Gauss methods   99
8.2 Diagonally implicit Runge-Kutta methods   100
    8.2.1 General format   100
    8.2.2 Specific methods   101
8.3 Other symplectic Runge-Kutta methods   102
8.4 Explicit Partitioned Runge-Kutta methods   103
    8.4.1 General format   103
    8.4.2 An alternative format   104
    8.4.3 Specific methods: orders 1 and 2   105
    8.4.4 Specific methods: order 3   107
    8.4.5 Specific methods: order 4 out of order 3   108
    8.4.6 Specific methods: order 4   109
    8.4.7 Specific methods: concatenations of the leap-frog methods   110
8.5 Available symplectic Runge-Kutta-Nyström methods   110
    8.5.1 Implicit methods   110
    8.5.2 Explicit methods: general format   110
    8.5.3 Specific explicit methods   111

9 Numerical experiments   **115**
9.1 A comparison of symplectic integrators   115
    9.1.1 Methods being compared   115
    9.1.2 Results: Kepler's problem   117
    9.1.3 Results: Hénon-Heiles problem   121
    9.1.4 Results: computation of frequencies   123
9.2 Variable step sizes for symplectic methods   124
9.3 Conclusions and recommendations   127

10 Properties of symplectic integrators — 129
- 10.1 Backward error interpretation — 129
  - 10.1.1 An example — 129
  - 10.1.2 The general case — 132
  - 10.1.3 Application to variable-step sizes — 133
- 10.2 An alternative approach — 134
- 10.3 Conservation of energy — 136
  - 10.3.1 Exact conservation: positive results — 136
  - 10.3.2 Exact conservation: negative results — 137
  - 10.3.3 Approximate conservation — 139
- 10.4 KAM theory — 141

11 Generating functions — 143
- 11.1 The concept of generating function — 143
  - 11.1.1 Introduction — 143
  - 11.1.2 Generating functions of the first kind — 143
  - 11.1.3 Generating functions of the third kind — 144
  - 11.1.4 Generating functions of all kinds — 145
- 11.2 Hamilton-Jacobi equations — 146
- 11.3 Integrators based on generating functions — 147
- 11.4 Generating functions for Runge-Kutta methods — 149
- 11.5 Canonical order theory — 150
  - 11.5.1 General framework — 150
  - 11.5.2 The Partitioned Runge-Kutta case — 151
  - 11.5.3 Elementary Hamiltonians — 153

12 Lie formalism — 155
- 12.1 The Poisson bracket — 155
- 12.2 Lie operators and Lie series — 156
  - 12.2.1 Lie operators — 156
  - 12.2.2 The adjoint representation — 157
  - 12.2.3 Lie series — 157
  - 12.2.4 Multiplication of exponentials — 159
- 12.3 The Baker-Campbell-Hausdorff formula — 160
- 12.4 Application to fractional-step methods — 161
  - 12.4.1 Introduction — 161
  - 12.4.2 The simplest splitting — 163
  - 12.4.3 Second-order splitting — 163
- 12.5 Extension to the non-Hamiltonian case — 164

13 High-order methods — 165
- 13.1 High-order Lie methods — 165

CONTENTS

13.1.1 Introduction   165
13.1.2 Yoshida's first approach: order 4   165
13.1.3 Yoshida's first approach: order 2r   167
13.1.4 Existence of symplectic methods of arbitrarily high orders   167
13.1.5 Yoshida's second approach   167
13.2 High-order Runge-Kutta-Nyström methods   170
    13.2.1 Order 7 methods   170
    13.2.2 Order 8 out of order 7   171
    13.2.3 Connection with the Lie formalism   172
13.3 A comparison of order 8 symplectic integrators   173
    13.3.1 Methods being compared   173
    13.3.2 Results: Kepler's problem   174
    13.3.3 Results: Hénon-Heiles problem   176
    13.3.4 Results: computation of frequencies   176
    13.3.5 Conclusions   177

14 Extensions   **179**
14.1 Partitioned Runge-Kutta methods for nonseparable Hamiltonian systems   179
14.2 Canonical B-series   180
14.3 Conjugate symplectic methods. Trapezoidal rule   181
14.4 Constrained systems   182
14.5 General Poisson structures   184
14.6 Multistep methods   185
14.7 Partial differential equations   186
14.8 Reversible systems. Volume-preserving flows   188

References   **189**

Symbol index   **197**

Index   **201**

Preface

'Niño, niño – dijo con voz alta a esta sazón don Quijote – seguid vuestra historia en línea recta, y no os metáis en las curvas y transversales.' M. de Cervantes, *El Ingenioso Hidalgo Don Quijote de la Mancha*, Parte II, Capítulo XXVI.

'Pray don't trouble yourself to say it any longer than that.' L. Carroll, *Alice's Adventures in Wonderland*, Chapter IX.

Recent years have witnessed a dramatic growth of the literature on symplectic integration of Hamiltonian problems. While the subject is still changing rapidly and important discoveries may yet be made, we feel it is time to present a unified view of this interdisciplinary field. The purpose of this book is to offer such a unified first introduction. Being exhaustive in the topics included and saying the last word on every issue treated have not been amongst our aims.

Some readers may be interested in integrating the Hamiltonian problems they find in their own scientific field. This sort of reader cannot reasonably be expected to be an expert in numerical methods. On the other hand, readers with an expertise in numerical methods may wish to enter the Hamiltonian field in order to design and analyse new Hamiltonian integrators. In our experience, readers in this second group are likely to be uncomfortable with the basic ideas of the Hamiltonian formalism. To cater for an audience as wide as possible, the book has five introductory chapters: two on Hamiltonian formalism and three on numerical methods. The main body of the book consists of Chapters 6 to 10. The four final chapters contain more advanced material. The book ends with a symbol index and an index.

Writing a book is a long task that cannot possibly be completed without help from many sources. In the particular case of this monograph, whose authors work in a teaching-oriented university with huge teaching loads, the value of the encouragement and assistance received from colleagues around the world cannot be overemphasized. Special thanks are due to A. Iserles and K.W. Morton (series editor) for the initial stimulus to write the book and to C. Grebogi, E. Hairer, B. Herbst, R. Skeel and G. Wanner who have aided us in various ways. We are also grateful to all our colleagues at the Department of Applied Mathematics and Computation in Valladolid and to the Spanish taxpayers, who financed this research through project PB89-0351 DGCYT. J.M.S. acknowledges the contribution of Mercedes, Carlos and Daniel. Without them the book would have been produced more quickly but less happily. M.P.C., on her part, acknowledges the support and encouragement received from Siro, Paz, Mar y Jose.

CHAPTER 1

Hamiltonian systems

1.1 Hamiltonian systems

This chapter and the next are a first introduction to Hamiltonian problems: more advanced material is presented later as required. A good starting point for the mathematical theory of Hamiltonian systems is the textbook by Arnold (1989). MacKay and Meiss (1987) have compiled an excellent collection of important papers in Hamiltonian dynamics. The article by Berry in this collection is particularly recommended. For an introduction to the more geometric modern approach the book by Marsden (1992) is an advisable choice.

We start by describing the class of problems we shall be concerned with and by presenting some notation. Let Ω be a domain (i.e., a nonempty, open, connected subset) in the oriented Euclidean space $\mathcal{R}^{2d}$ of the points $(\mathbf{p}, \mathbf{q}) = (p_1, \ldots, p_d, q_1, \ldots, q_d)$. We denote by I an open interval of the real line $\mathcal{R}$ of the variable t (time); I may be bounded, $I = (a, b)$, or unbounded, $I = (-\infty, b)$, $I = (a, \infty)$, $I = (-\infty, \infty)$. If $H = H(\mathbf{p}, \mathbf{q}, t)$ is a sufficiently smooth real function defined in the product $\Omega \times I$, then the *Hamiltonian system* of differential equations with Hamiltonian H is, by definition, given by

$$\frac{dp_i}{dt} = -\frac{\partial H}{\partial q_i}, \quad \frac{dq_i}{dt} = +\frac{\partial H}{\partial p_i}, \quad i = 1, \ldots, d. \qquad (1.1)$$

The integer d is called the *number of degrees of freedom* and Ω is the *phase space*. The product $\Omega \times I$ is the *extended phase space*. The exact amount of smoothness required of H will vary from place to place and will not be explicitly stated, but throughout we assume at least C^2 continuity, so that the right-hand side of the system (1.1) is C^1 and the standard existence and uniqueness theorems apply to the corresponding initial value problem. Sometimes, the symbol S_H will be used to refer to the system (1.1).

Usually, in applications to mechanics (Arnold (1989)), the **q** variables are *generalized coordinates*, the **p** variables the conjugated *generalized momenta* and H corresponds to the total *mechanical energy*.

In many Hamiltonian systems of interest, the Hamiltonian H does not explicitly depend on t; then (1.1) is an *autonomous* system of differential equations. For autonomous problems we shall consider H as a function defined in the phase space Ω, rather than as a function defined in $\Omega \times \mathcal{R}$ and independent of the last variable.

It is sometimes useful to combine all the dependent variables in (1.1) in a $2d$-dimensional vector $\mathbf{y} = (\mathbf{p}, \mathbf{q})$. Then (1.1) takes the simple form

$$\frac{d\mathbf{y}}{dt} = J^{-1}\nabla H, \qquad (1.2)$$

where ∇ is the *gradient* operator

$$(\partial/\partial p_1, \ldots, \partial/\partial p_d, \partial/\partial q_1, \ldots, \partial/\partial q_d) \qquad (1.3)$$

and J is the $2d \times 2d$ skew-symmetric matrix

$$J = \begin{bmatrix} 0 & I \\ -I & 0 \end{bmatrix} \qquad (1.4)$$

(I and 0 respectively represent the unit and zero $d \times d$ matrices).

Upon differentiation of H with respect to t along a solution of (1.1), we find

$$\frac{d}{dt} H(\mathbf{y}(t), t) = (\nabla H)^T \frac{d\mathbf{y}}{dt} + \frac{\partial H}{\partial t},$$

so that, in view of (1.2) and of the skew-symmetry of J^{-1},

$$\frac{dH}{dt} = (\nabla H)^T J^{-1} \nabla H + \frac{\partial H}{\partial t} = \frac{\partial H}{\partial t}.$$

In particular, if H is autonomous, $dH/dt = 0$. Then H is a conserved quantity that remains constant along solutions of the system. In the applications, this usually corresponds to *conservation of energy*.

We now turn to some concrete examples of Hamiltonian systems. These examples have been chosen for their simplicity. More realistic examples from celestial mechanics, plasma physics, molecular dynamics etc. can be found in the literature of the corresponding fields.

1.2 Examples of Hamiltonian systems

1.2.1 The harmonic oscillator

This is the well-known system with $d = 1$ (one degree of freedom)
$$H = T + V, \quad T = p_1^2/(2m), \quad V = kq_1^2/2.$$
Here, m and k are positive constants that, for the familiar case of a material point attached to a spring, respectively correspond to mass and spring constant. Of course T and V are the *kinetic* and *potential* energies.

In situations with $d = 1$, it is clearly convenient to use the notation p and q for the dependent variables (rather than p_1, q_1). With this notation, the equations (1.1) for the harmonic oscillator read
$$\dot{p} = -kq, \quad \dot{q} = \frac{p}{m}. \tag{1.5}$$
Here and elsewhere dots represent differentiation with respect to t. The general solution for q is an oscillation
$$q(t) = C_1 \sin(\omega t + C_2), \quad \omega = \sqrt{(k/m)},$$
with angular frequency ω (period $T = 2\pi/\omega$, frequency $\nu = 1/T = \omega/(2\pi)$); C_1 and C_2 are integration constants. Similarly p is given by
$$p(t) = m\omega C_1 \cos(\omega t + C_2).$$
The particular solution that takes the initial value $(p(0), q(0))$ at $t = 0$ is easily written in matrix form:
$$\begin{bmatrix} p(t) \\ q(t) \end{bmatrix} = \begin{bmatrix} \cos \omega t & -(m\omega) \sin \omega t \\ (m\omega)^{-1} \sin \omega t & \cos \omega t \end{bmatrix} \begin{bmatrix} p(0) \\ q(0) \end{bmatrix}. \tag{1.6}$$
When plotted in the phase (p, q)-plane, the parametric curves $(p(t), q(t))$ correspond to the *ellipses*
$$H(p, q) = p^2/(2m) + kq^2/2 = \text{constant}.$$
These are circles when $mk = 1$ (or, equivalently, when $m\omega = 1$).

1.2.2 The pendulum

If the units are chosen in such a way that the mass of the blob, the length of the rod and the acceleration of gravity are all unity, then
$$H = T + V, \quad T = p^2/2, \quad V = -\cos q,$$

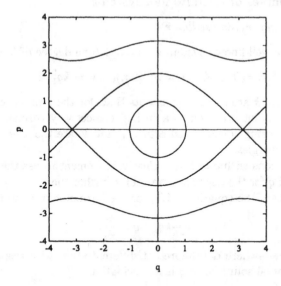

Figure 1.1. *Phase plane of the pendulum*

where q is the angle between the rod and a vertical, downward oriented axis. The equations of motion are then

$$\dot{p} = -\sin q, \quad \dot{q} = p.$$

In the phase (p, q)-plane depicted in Fig. 1.1 the solutions lie in the level curves $H = constant$. We only consider the situation near $q = 0$; the phase portrait repeats itself periodically along the q-axis because H is a periodic function of q. There is a stable equilibrium at $p = 0$, $q = 0$ (the pendulum rests in its lowest position), surrounded by *libration* solutions where q varies periodically between values $q_{max} > 0$ and $q_{min} = -q_{max}$. The libration trajectories fill the region $-1 < H < 1$. For $H > 1$ we find *rotation* solutions, where q varies monotonically. The level set $H = 1$ is composed of the unstable equilibria at $q = \pi$, $q = -\pi$ (pendulum resting in the highest position) and of the *separatrices* connecting them.

To integrate the equations of motion, we substitute p by $\dot{q}$ in the energy equation $H(p, q) = h$, with h a constant. This yields a differential equation

$$dt = \frac{dq}{\sqrt{2(h - V(q))}}, \tag{1.7}$$

that is readily integrated in terms of quadratures of elementary functions.

In particular, we see that the period of a libration solution is given by

$$T = 2\int_{-q_{max}}^{q_{max}} \frac{dq}{\sqrt{2(h - V(q))}}.$$

Note that $h = H(0, q_{max}) = -\cos q_{max}$. For amplitudes q_{max} close to 0, the period T is close to 2π (the period of the linearized equations $\dot{p} = -q$, $\dot{q} = p$). As q_{max} approaches π, the trajectory approaches the separatrix and the period approaches ∞. The dependence of the period on the amplitude is typical of *nonlinear* oscillations; for the (linear) harmonic oscillator all solutions have of course the same period.

1.2.3 The double harmonic oscillator

This has two degrees of freedom and

$$H = T + V, \quad T = (\omega_1 p_1^2 + \omega_2 p_2^2)/2, \quad V(q) = (\omega_1 q_1^2 + \omega_2 q_2^2)/2.$$

In the equations of motion $\dot{p}_1 = -\omega_1 q_1$, $\dot{q}_1 = \omega_1 p_1$, $\dot{p}_2 = -\omega_2 q_2$, $\dot{q}_2 = \omega_2 p_2$, the (p_1, q_1) variables are not coupled to the variables (p_2, q_2); we are considering two uncoupled harmonic oscillators. According to our previous discussion of the harmonic oscillator, the projections of the solutions of the double harmonic oscillator onto the (p_i, q_i)-plane correspond to circles and possess angular frequency $\omega_i > 0$. If ω_1/ω_2 is a rational number r/s, then the solutions of the double harmonic oscillator are periodic, with period $T = 2\pi r/\omega_1 = 2\pi s/\omega_2$; the trajectory returns to its initial position in the four-dimensional phase space after having completed r cycles of the (p_1, q_1) variables and s cycles of the (p_2, q_2) variables. If ω_1/ω_2 is irrational, the trajectory never returns to its initial location.

A geometric picture is useful. The system has two conserved quantities

$$H_1 = \frac{\omega_1}{2}(p_1^2 + q_1^2), \quad H_2 = \frac{\omega_2}{2}(p_2^2 + q_2^2),$$

that represent the energies in each of the two uncoupled oscillators. In the phase space (p_1, p_2, q_1, q_2), the level sets $H_1 = constant_1$, $H_2 = constant_2$ represent 2-dimensional tori. These tori are *invariant*: if a trajectory is at time $t = 0$ on one of the tori it is on that torus for all times t. The phase $\phi_1 = \arctan(p_1/q_1)$ of the first oscillation represents the *longitude* in the torus and the phase ϕ_2 of the second oscillation represents the *latitude* in the torus. The longitude and latitude along a trajectory vary periodically with angular frequencies ω_1 and ω_2. For $\omega_1/\omega_2 = r/s$ the trajectory returns to its initial location after winding itself on the torus r times in the direction of the parallels and s times in the direction of the meridians. For ω_1/ω_2 irrational, the trajectory never returns to its initial position. It can be shown that it is actually dense on the torus surface, and even *ergodic*, i.e., the trajectory stays in each domain D on the torus surface an amount of time proportional to the area of D (Arnold (1989), Section 51).

In the case with ω_1/ω_2 irrational the vector-valued function $(p_1(t), p_2(t), q_1(t), q_2(t))$ is *quasiperiodic* (Siegel and Moser (1971), Section 36). In general, a function $\mathbf{F}(t)$ is said to be quasiperiodic, with frequencies ω_1, ω_2, if it can be expanded in a series of the form

$$\sum_{m,n=-\infty}^{\infty} [\mathbf{a}_{m,n} \cos(m\omega_1 + n\omega_2)t + \mathbf{b}_{m,n} \sin(m\omega_1 + n\omega_2)t].$$

If ω_2 is an integer multiple of ω_1 (or, more generally, if ω_1/ω_2 is rational), then this series reduces to a Fourier series for a periodic function. All quantities $m\omega_1 + n\omega_2$ are then integer multiples of a single value ω.

Quasiperiodic functions with $k > 2$ frequencies ω_i can be defined in an obvious way and would appear in the study of k uncoupled harmonic oscillators.

1.2.4 Kepler's problem

Kepler's problem describes the motion in a plane (the *configuration plane*) of a material point that is attracted towards the origin with a force inversely proportional to the distance squared. In nondimensional form,

$$H = T + V, \quad T = (1/2)(p_1^2 + p_2^2), \quad V = -1/\sqrt{q_1^2 + q_2^2}.$$

The equations of motion are then

$$\dot{p}_i = -\frac{q_i}{(q_1^2 + q_2^2)^{3/2}}, \quad \dot{q}_i = p_i, \quad i = 1, 2.$$

Since the problem is autonomous, the Hamiltonian (energy) H is a conserved quantity. Furthermore, due to the *central* character of

EXAMPLES OF HAMILTONIAN SYSTEMS

the force (Arnold (1989), Sections 6–7), there is a second conserved quantity: the *angular momentum*

$$M = q_1 p_2 - q_2 p_1. \tag{1.8}$$

For the analysis, it is best to employ polar coordinates (r, θ) in the configuration (q_1, q_2)-plane. Then, the corresponding momenta are $p_r = \dot{r}$ and $p_\theta = r^2 \dot{\theta}$ and the Hamiltonian becomes

$$H = T + V, \quad T = \frac{1}{2}\left(p_r^2 + \frac{p_\theta^2}{r^2}\right), \quad V = -\frac{1}{r}.$$

The equations (1.1) are then

$$\dot{p}_r = \frac{p_\theta^2}{r^3} - V'(r), \tag{1.9}$$

$$\dot{p}_\theta = 0, \tag{1.10}$$

$$\dot{r} = p_r, \tag{1.11}$$

$$\dot{\theta} = \frac{p_\theta}{r^2}. \tag{1.12}$$

From (1.10) we see that p_θ is a constant of motion; in fact an easy computation shows that p_θ is in fact the polar coordinate expression of the angular momentum M whose cartesian expression is (1.8). Upon replacing p_θ by a constant M and p_r by $\dot{r}$ in the equation $H = h = constant$, that expresses conservation of energy, we obtain a first-order differential equation for r

$$dt = \frac{dr}{\sqrt{2(h - (M^2/2r^2) - V(r))}} \tag{1.13}$$

that can be solved by quadratures, cf. (1.7). If the constant H is negative and $M \neq 0$, then $r = r(t)$ librates periodically between a minimum $r_{min} > 0$ and a maximum r_{max}. The points in the configuration plane where r is minimum are called *pericentres;* those corresponding to maximum r are called *apocentres*. The period of r is found to be (Arnold (1989), Section 8)

$$\mathcal{T} = \frac{2\pi}{\sqrt{2|H|}^3}. \tag{1.14}$$

Once $r = r(t)$ is known, a quadrature in (1.12) (with $p_\theta = M$) yields $\theta = \theta(t)$. It turns out that the polar angle θ between a pericentre and the next pericentre is exactly 2π. Hence after time $\mathcal{T}$, not only r reassumes its initial value, but the moving material point returns to its initial position in the configuration plane.

Hence the trajectory in this plane is a closed curve; this trajectory is an ellipse, of course. All four functions $(p_r, p_\theta, r, \theta)$ (or the cartesian (p_1, p_2, q_1, q_2)) are periodic with period (1.14).

Example 1.1 Consider the initial conditions

$$p_1 = 0, \quad p_2 = \sqrt{\frac{1+e}{1-e}}, \quad q_1 = 1 - e, \quad q_2 = 0, \qquad (1.15)$$

where e is a parameter ($0 \leq e < 1$). The period (1.14) of the solution is readily found to be 2π. The values r_{max} and r_{min} can be computed by setting $\dot{r} = 0$ in the equation of conservation of energy. It turns out that $r_{max} = 1 + e$ and $r_{min} = 1 - e$. Hence the initial condition (1.15) corresponds to the pericentre and the major semiaxis of the ellipse is 1. Furthermore the distance from the centre of the ellipse to the origin (focus of the ellipse) equals e, so that the parameter e represents the *eccentricity*. □

1.2.5 A modified Kepler problem

In many applications the Kepler potential $V = -1/r$ has to be corrected in various ways. For instance (Kirchgraber (1988)), the Hamiltonian

$$H = T + V, \quad T = \frac{p_1^2 + p_2^2}{2}, \quad V = -\frac{1}{\sqrt{q_1^2 + q_2^2}} - \frac{\epsilon}{2\sqrt{(q_1^2 + q_2^2)^3}}$$

(ϵ is a small *perturbation* parameter) corresponds to the motion in a plane of a particle gravitationally attracted by a slightly oblate sphere (rather than by a point mass). The attracting body is rotationally symmetric with respect to an axis orthogonal to the plane of the particle. (This is the situation with an artificial satellite moving in the Earth's equatorial plane.)

The problem has the energy H and the angular momentum M as conserved quantities. For the analysis it is again best to use polar coordinates. The equations (1.9)–(1.12) are still valid; the only change is that now $V = -1/r - \epsilon/(2r^3)$. We proceed as in the unperturbed Kepler problem to arrive once more at (1.13). For $H < 0$, $M \neq 0$, the polar radius r librates between values r_{max} and r_{min} with a period $\mathcal{T} = \mathcal{T}(\epsilon)$ (of course $\mathcal{T}(0)$ coincides with (1.14)).

The main difference with the unperturbed Kepler problem is that now, after integrating (1.12), the polar angle between a pericentre and the next is not exactly 2π (it differs from 2π in an $O(\epsilon)$ quantity). As a consequence the orbit in the (q_1, q_2)-plane

EXAMPLES OF HAMILTONIAN SYSTEMS

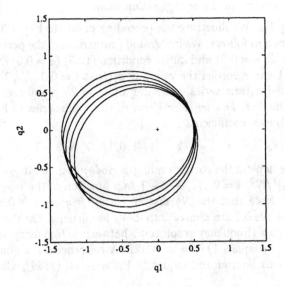

Figure 1.2. *Precession of the pericentre in the perturbed Kepler problem*

does not quite close after $T(\epsilon)$ units of time. This is illustrated in Fig. 1.2, where $\epsilon = 0.01$, the initial condition is given by (1.15) ($e = 0.5$) and $0 \leq t \leq 8\pi$. For the unperturbed potential, the moving particle would have returned to the initial position after completing four revolutions around the origin. Here we observe the phenomenon of *precession* of the pericentre. Now the solution is in general *quasiperiodic*, rather than periodic as in the unperturbed Kepler problem.

In order to better understand the quasiperiodic character of the solutions, we note that, in the four-dimensional phase space, the surfaces $H = constant$, $M = constant$ are topologically tori. The *latitude* angle on a torus corresponds to the phase of the libration of the r and p_r variables; the *longitude* to the polar angle θ. In the unperturbed potential ($\epsilon = 0$), the period of the r-libration exactly equals the time needed for the trajectory to go once around the origin in the configuration (q_1, q_2)-plane (i.e., the time needed by θ to increase by 2π). When the perturbation is present, the frequencies of the latitude and longitude are slightly different; their difference ν_p is the frequency of the precession of the pericentre, i.e., the number of times per unit of time that the pericentre rotates around the origin in the configuration plane.

Example 1.2 We illustrate the preceding points in Fig. 1.3, that was obtained as follows. We integrated (numerically) the perturbed problem with $\epsilon = 0.01$ and initial condition (1.15) ($e = 0.5$) for $0 \le t \le 2^{13} - 1$ and recorded the values of $q_1(t)$ at $t = 0, 1, \ldots, 2^{13} - 1$. This yielded a time series consisting of $N = 2^{13}$ numbers. Then we took the *Fast Fourier Transform* of this time series to find N discrete Fourier coefficients

$$Q_i, \quad i = -N/2, \ldots, -1, 0, 1, \ldots, N/2 - 1.$$

The figure depicts the corresponding *periodogram*, i.e., it gives the values $|Q_i|^2/N^2$, $i = 0, \ldots, N/2 - 1$ as a function of the frequency $\nu_i = i/N$. Note that the *Nyquist or cut-off frequency* is 0.5; frequencies above 0.5 are *aliased* into lower frequencies. On the other hand the gap (frequency resolution) between a frequency ν_i and the next ν_{i+1} equals $1/N = 0.000122$. A reference for a quick introduction to Fourier techniques is Press *et al.* (1989), Chapter 12.

We see that in the function $q_1(t)$ there are essentially two frequencies. The fastest equals 0.1837 (approximately $1/(2\pi)$) and corresponds to the rotation of the moving point around the origin in the configuration plane. The slowest 0.0051 represents the frequency of precession of the pericentre. There is a smaller spectral line at $\nu = 0.3623$; since 0.3623 equals $2 \times 0.1837 - 0.0051$, this third frequency is a combination with integer coefficients of the two main frequencies. This is consistent with the quasiperiodicity of the solutions. □

There is yet another way of displaying the quasiperiodicity of the solutions. For constant h, the level set S_h defined in phase space by $H = h$ is three-dimensional. Each solution of the equations of motion remains in a set S_h. In S_h we can see (p_1, q_1, q_2) as independent variables; the value of p_2 can be retrieved (up to the sign) from the values of (p_1, q_1, q_2) by using the equation $H = h$. We intersect S_h set with the plane $q_2 = 0$ to obtain a *Poincaré section*. In the section only (p_1, q_1) are independent variables. This makes it possible to represent on a (two-dimensional) plane solutions of four-dimensional differential equations. Given a solution of the equations of motion we plot in the (p_1, q_1) section-plane the points where this solution intersects the $q_2 = 0$ plane.

Example 1.3 In Fig. 1.4 we have done this for the case $\epsilon = 0.01$ with initial condition (1.15) ($e = 0.5$). The first 75 intersections

EXAMPLES OF HAMILTONIAN SYSTEMS

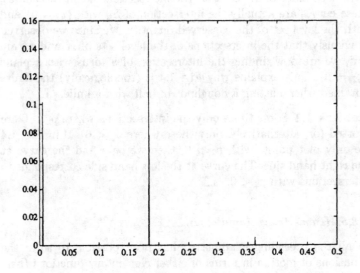

Figure 1.3. *Spectrum of a modified Kepler solution*

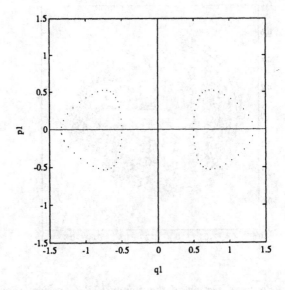

Figure 1.4. *Poincaré section for the modified Kepler problem*

have been plotted. We see that the intersections lie on two curves; these curves are actually the intersection of S_h with $\{q_2 = 0\}$ and with the level set of the conserved quantity M. Since we observed previously that the intersections of the level sets of H and M are tori, we are now finding the intersection of a torus with a plane $\{q_2 = 0\}$. This explains why Fig. 1.4 is (topologically) the figure obtained when cutting a doughnut in half with a knife. □

Remark 1.1 Some times only the intersections where $p_2 > 0$ are plotted (or, alternatively, only those where $p_2 < 0$). If in Fig. 1.4 we only plot points with $p_2 > 0$, then we only find the curve at the right-hand side. The curve at the left-hand side corresponds to intersections with $p_2 < 0$. □

1.2.6 Hénon-Heiles Hamiltonian

In all the preceding examples it has been possible to solve the equations of motion in terms of either elementary functions (harmonic oscillator and double harmonic oscillator) or in terms of quadratures of elementary functions (pendulum, Kepler and mod-

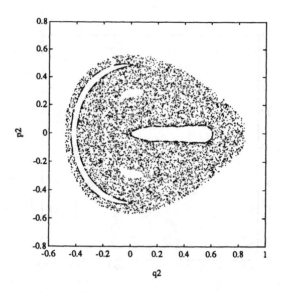

Figure 1.5. *Poincaré section of the Hénon-Heiles problem showing a chaotic solution*

EXAMPLES OF HAMILTONIAN SYSTEMS

ified Kepler problems). In the case of the harmonic oscillator and double harmonic oscillator the differential equations are linear with constant coefficients and the integration is trivial. In the three nonlinear examples the integration has been possible through the use of the conserved quantities (H for the pendulum and H and M for the Kepler and modified Kepler problems). In general and without being too precise, a Hamiltonian system with d degrees of freedom for which d independent conserved quantities exist can be integrated by quadratures (Arnold (1989), Section 49).

However not all Hamiltonian systems are *integrable*. Hénon and Heiles (1964) constructed the celebrated Hamiltonian

$$H = T + V, \quad T = \frac{1}{2}(p_1^2 + p_2^2), \quad V = \frac{1}{2}(q_1^2 + q_2^2) + q_1^2 q_2 - \frac{1}{3}q_2^2.$$

In Fig. 1.5 we have displayed the (p_2, q_2) Poincaré section with the $q_1 = 0$ plane. The initial condition has $q_1 = q_2 = p_2 = 0$ with p_1 determined from the energy equation so as to have $H = 0.15925$. The integration is performed numerically for $0 \leq t \leq 30000$. We see that the intersections depicted, corresponding to one and the same trajectory, do not stay on a curve, but rather are

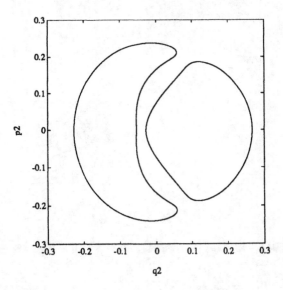

Figure 1.6. *Poincaré section of the Hénon-Heiles problem showing a quasiperiodic solution*

'randomly' scattered in a two-dimensional region. This proves that a conserved quantity F independent of H cannot exist, for if such an F existed the intersections would lie on a curve $H = constant$, $F = constant$, $q_1 = 0$. Trajectories like that in Fig. 1.5 are called *chaotic* or *stochastic*.

Fig. 1.6 is again a (p_2, q_2) Poincaré section, but now the initial condition is $p_1 = p_2 = q_1 = q_2 = 0.12$ and the integration time $0 \leq t \leq 200000$. The intersections do lie on a curve suggesting a quasiperiodic behaviour of the solution and the presence of an individual invariant torus.

Hamiltonian systems are typically nonintegrable and have simultaneously stochastic and quasiperiodic solutions. The phase space is then composed of invariant tori and gaps between them filled by stochastic orbits. This is, to some extent, explained by the *KAM (Kolmogorov-Arnold-Moser) theory* (see the article by Berry in the book by MacKay and Meiss (1987)).

CHAPTER 2

Symplecticness

2.1 The solution operator

Our goal in this chapter is the introduction of the property of symplecticness, which plays a very important role throughout the book. Symplecticness is a characterization of Hamiltonian systems in terms of their *solutions*, rather than in terms of the actual form of the differential equations. We first need the concept of *solution operator*.

If t, t_0 are real numbers in the interval I, then we denote by $\Phi_H(t, t_0)$ the solution operator of the system (1.1). By definition $\Phi_H(t, t_0)$ is a transformation mapping the phase space Ω into itself, in such a way that, for $(\mathbf{p}^0, \mathbf{q}^0)$ in Ω,

$$(\mathbf{p}, \mathbf{q}) = \Phi_H(t, t_0)(\mathbf{p}^0, \mathbf{q}^0) \qquad (2.1)$$

is the value at time t of the solution of (1.1) that at time t_0 has the initial condition $(\mathbf{p}^0, \mathbf{q}^0)$. Thus, if in (2.1) t varies and t_0, $(\mathbf{p}^0, \mathbf{q}^0)$ are seen as fixed, then we recover the solution of (1.1) with initial condition $(\mathbf{p}^0, \mathbf{q}^0)$ at time t_0. The key point is that we will mainly be interested in seeing t, t_0 in (2.1) as fixed parameters and $(\mathbf{p}^0, \mathbf{q}^0)$ as a variable, so that we are defining a transformation of Ω into itself.

The transformed point $\Phi_H(t, t_0)(\mathbf{p}^0, \mathbf{q}^0)$ is defined only if the solution of $\mathcal{S}_H$ with initial condition $(\mathbf{p}^0, \mathbf{q}^0)$ at t_0 exists at time t, which, for given $(\mathbf{p}^0, \mathbf{q}^0)$ is not necessarily the case if $|t - t_0|$ is large: solutions may either reach the boundary of Ω or go to ∞ in a finite time and exist only for bounded time intervals. Thus for given t and t_0 the domain of $\Phi_H(t, t_0)$ may be strictly smaller than Ω.

Example 2.1 For the autonomous, one-degree-of-freedom Hamiltonian $H = pq^2$, with equations of motion

$$\dot{p} = -2pq, \quad \dot{q} = q^2,$$

an elementary integration shows that

$$\Phi_H(t,t_0)(p^0,q^0) = \left(p^0(1-(t-t_0)q^0)^2, \frac{q^0}{1-(t-t_0)q^0}\right).$$

The domain of $\Phi_H(t,t_0)$, $t - t_0 > 0$, is the set

$$\{q^0 < 1/(t-t_0)\},$$

in spite of the fact that H is defined and smooth in the whole of the phase space $\mathcal{R}^2$. $\square$

The operators $\Phi_H(t,t_0)$ satisfy the equation

$$\Phi_H(t_2,t_0) = \Phi_H(t_2,t_1)\Phi_H(t_1,t_0), \qquad t_2,t_1,t_0 \in I. \qquad (2.2)$$

This equality means that the domain of $\Phi_H(t_2,t_0)$ is the same as the domain where the composition $\Phi_H(t_2,t_1)\Phi_H(t_1,t_0)$ is defined, and that both sides of (2.2) coincide in their common domain. Some readers may wish to check the validity of (2.2) for the Hamiltonian in Example 2.1.

For autonomous Hamiltonians, $\Phi_H(t,t_0)$ only depends on the difference $t - t_0$ (see e.g. Example 2.1). We then write $\phi_{t-t_0,H}$ instead of $\Phi_H(t,t_0)$. The transformation in phase space given by $\phi_{s,H}$ is called the *s-flow* of the system S_H. For flows, (2.2) becomes the *semigroup property*

$$\phi_{t+s,H} = \phi_{t,H}\phi_{s,H}. \qquad (2.3)$$

Again, this relation implies the coincidence of the domains of the left- and right-hand sides.

2.2 Preservation of area by one-degree-of-freedom Hamiltonian systems

2.2.1 Concept of preservation of area

The idea of symplectic integration revolves around the use of symplectic transformations. In our experience the notion of symplecticness is likely to cause confusion when first met. It is therefore important that we devote some time to understanding symplecticness. It is best to start with the one-degree-of-freedom case, where symplecticness is nothing but preservation of (oriented) area. We then assume, until further notice, that $d = 1$.

For each real t, t_0 the solution mapping $\Phi_H(t,t_0)$ is an *area-preserving* transformation in Ω, in the sense that, for each bounded subdomain $D \subset \Omega$ for which $\Phi_H(t,t_0)(D)$ is defined, it holds true

PRESERVATION OF AREA

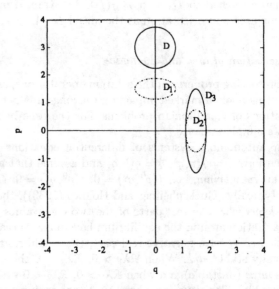

Figure 2.1. *Preservation of area by the flow of the harmonic oscillator*

that D and $\Phi_H(t,t_0)(D)$ have the same area (and orientation). To see this, it is enough, after recalling Liouville's theorem (see e.g. Section 3.5, Chapter 1 in the article by Arnold and Il'yashenko in Anosov and Arnold (1988)), to observe that, for each t, the vector field in phase space $[-\partial H/\partial q, \partial H/\partial p]^T$ that features in (1.1) is divergence-free because

$$\frac{\partial}{\partial p}\left(-\frac{\partial H}{\partial q}\right) + \frac{\partial}{\partial q}\left(\frac{\partial H}{\partial p}\right) = 0.$$

Example 2.2 For the harmonic oscillator (1.5) with $m\omega = 1$ the area-preserving property is evident: in this case the t-flow (1.6) is a rigid rotation by ωt radians. To check this property for $m\omega \neq 1$, note that the matrix in (1.6) can be factorized as

$$\begin{bmatrix} m\omega & 0 \\ 0 & 1 \end{bmatrix} \begin{bmatrix} \cos\omega t & -\sin\omega t \\ \sin\omega t & \cos\omega t \end{bmatrix} \begin{bmatrix} (m\omega)^{-1} & 0 \\ 0 & 1 \end{bmatrix}.$$

The rightmost matrix contracts areas by a factor $m\omega$, an effect that is just offset by the leftmost matrix that operates after the rigid rotation. An illustration is provided in Figure 2.1, where $\omega = 1$, $t = \pi/2$, $m\omega = 2$. The circular domain D is mapped by the rightmost transformation into the ellipse D_1, which is rotated to

D_2 and then stretched to $D_3 = \phi_{\pi/2,H}(D)$. The transformed D_3 has the same area and orientation as the initial D. □

2.2.2 Preservation of area and dynamics

The area-preserving property of the solution operator has, as studied later in this book, a marked impact on the long-time behaviour of the solutions of Hamiltonian problems. For the time being, let us present a simple instance.

Consider autonomous systems of differential equations in the plane of the form $\dot{p} = f(p,q)$, $\dot{q} = g(p,q)$ and assume that a point (p^0, q^0) is an *equilibrium*, i.e., $f(p^0, q^0) = 0$, $g(p^0, q^0) = 0$. *Generically*, i.e., typically (Guckenheimer and Holmes (1983)), the equilibrium is *hyperbolic:* the real parts of the two eigenvalues λ_1, λ_2 of the linearization around the equilibrium have nonzero real part. When both λ_1, λ_2 have negative real part, the equilibrium is an asymptotically stable *sink*. When $\Re\lambda_1 > 0$, $\Re\lambda_2 > 0$ the equilibrium is a *source* (unstable) and when $\Re\lambda_1 > 0$, $\Re\lambda_2 < 0$ we have a *saddle* (unstable). The situation where λ_1, λ_2 are conjugate purely imaginary complex numbers does not arise generically: if a system is in that case, then there are arbitrarily small perturbations of the right-hand side that turn the system into one with an equilibrium which is either a sink or a source. However, if we restrict our attention to (autonomous) *Hamiltonian* problems in the plane the situation changes completely: sinks (or sources) cannot occur, because in their neighbourhood the flow contracts (expands) areas. The case $\Re\lambda_1 = \Re\lambda_2 = 0$ is now generic: if a Hamiltonian system is in that case *(centre)*, then, typically, neighbouring Hamiltonian systems are in the same case.

Considerations similar to those just made for equilibria apply to *periodic solutions*. In 'general' systems of autonomous ordinary differential equations in the plane, typical periodic orbits are stable or unstable *limit cycles*. Near such limit cycles area contracts or expands and hence limit cycles cannot appear at all in the Hamiltonian case. On the other hand, for autonomous Hamiltonian systems in the plane, a periodic orbit is typically surrounded by periodic orbits (as in the harmonic oscillator or pendulum examples), a situation that is not generic in the non-Hamiltonian case.

We see that, via the conservation of area property, features that are exceptional for general systems become the rule for Hamiltonian systems. Conversely, features that are typical for general systems cannot arise in Hamiltonian problems.

2.2.3 Preservation of area as a characteristic property

It turns out that all properties specific to the Hamiltonian dynamics can be derived from the preservation of area property. This is no surprise because the area-preserving character of the solution operator, which was shown above to hold for Hamiltonian systems, actually holds *only* for Hamiltonian systems. More precisely, assume that Ω is simply connected, i.e., it has no holes, and suppose that

$$\frac{dp}{dt} = f(p,q,t), \quad \frac{dq}{dt} = g(p,q,t), \qquad (2.4)$$

is a smooth differential system whose solution operator is area-preserving. Then (2.4) is actually a Hamiltonian system S_H for a suitable H. There is nothing deep about this. By Liouville's theorem, for each fixed t, the vector field $[f,g]^T$ is divergence-free, so that

$$\frac{\partial}{\partial p}(-f) = \frac{\partial}{\partial q}g.$$

But this is just the necessary and sufficient condition for the field $[g, -f]^T$ to be the gradient of a scalar function H, i.e., for (2.4) to coincide with S_H.

Remark 2.1 If Ω is not simply connected, then systems with area-preserving solution operators are, in general, only *locally Hamiltonian:* in each ball $B \subset \Omega$ they coincide with a Hamiltonian system S_{H_B} but, globally, the system may not be Hamiltonian because the various H_B cannot be patched together. A typical example is given by the area-preserving system

$$\frac{dp}{dt} = \frac{p}{p^2+q^2}, \quad \frac{dq}{dt} = \frac{q}{p^2+q^2}$$

defined in $\Omega = \mathcal{R}^2 \backslash (0,0)$. In each ball in Ω the system is Hamiltonian with H given by a branch of the argument of the point (p,q). The system is not Hamiltonian because of course the argument cannot be defined as a smooth single-valued function in $\mathcal{R}^2 \backslash (0,0)$. □

2.3 Checking preservation of area: Jacobians

Let $(p^*, q^*) = \psi(p,q)$ be a $\mathcal{C}^1$ transformation defined in a domain Ω. According to the standard rule for changing variables in an integral, ψ preserves area and orientation if and only if the Jacobian determinant is identically 1:

$$\forall (p,q) \in \Omega, \quad \frac{\partial p^*}{\partial p}\frac{\partial q^*}{\partial q} - \frac{\partial p^*}{\partial q}\frac{\partial q^*}{\partial p} = 1. \qquad (2.5)$$

It is a trivial exercise in matrix multiplication to check that this relation can be rewritten as

$$\forall (p,q) \in \Omega, \quad \psi'^T J \psi' = \frac{\partial (p^*, q^*)}{\partial (p, q)}^T J \frac{\partial (p^*, q^*)}{\partial (p, q)} = J, \qquad (2.6)$$

where

$$\psi' = \frac{\partial (p^*, q^*)}{\partial (p, q)} = \begin{bmatrix} \partial p^*/\partial p & \partial p^*/\partial q \\ \partial q^*/\partial p & \partial q^*/\partial q \end{bmatrix}$$

is the Jacobian matrix of the transformation and J is the matrix in (1.4) (with $d=1$). Going from (2.5) to (2.6) may appear to be just a matter of complicating things. This is not so: the matrix J is a very important character in this play. *If* **v** *and* **w** *are vectors in the plane, then* $\mathbf{v}^T J \mathbf{w}$ *is the oriented area of the parallelogram they determine.* (The importance of J is the reason why we write the matrix in the Hamiltonian system (1.2) as J^{-1}, rather than using a specific symbol for J^{-1}.)

Now let us fix a point (p,q) in Ω and construct a parallelogram $\mathcal{P}$ having a vertex at (p,q) and having as sides two small vectors **v** and **w** (i.e., the vertices are the points (p,q), $(p,q)+\mathbf{v}$, $(p,q)+\mathbf{w}$, $(p,q)+\mathbf{v}+\mathbf{w}$). Then $\psi(\mathcal{P})$ is a parallelogram with curved sides, which can be approximated by the parallelogram $\mathcal{P}^*$ based at $\psi(p,q)$ with sides $\psi'\mathbf{v}$, $\psi'\mathbf{w}$. In fact, by the very definition of ψ', $\psi(\mathcal{P})$ and $\mathcal{P}^*$ differ in terms higher than linear in **v** and **w**. Now $\mathcal{P}^*$ and $\mathcal{P}$ have the same area if and only if

$$\mathbf{v}^T \psi'^T J \psi' \mathbf{w} = \mathbf{v}^T J \mathbf{w}.$$

Clearly, the last relation holds for all parallelograms $\mathcal{P}$ in Ω if and only if (2.6) holds. The conclusion is that (2.6) means that, at each point $(p,q) \in \Omega$, the linear transformation ψ' maps parallelograms based at (p,q) into parallelograms based at $\psi(p,q)$ without altering the oriented area. The conservation of area by ψ is then a consequence of the conservation of the area of 'infinitesimal' parallelograms implied in (2.6).

2.4 Checking preservation of area: differential forms

Differential forms in Ω provide an alternative language to express the considerations made in the preceding subsection. A detailed

SYMPLECTIC TRANSFORMATIONS   21

study of the meaning and properties of differential forms is definitely outside the scope of this book (the interested reader is referred to Arnold (1989), Chapter 7). However the algebraic manipulations required to prove conservation of area via differential forms are as a rule easier than those required to prove conservation of area via (2.6). It is therefore advisable to comment, albeit briefly, on differential forms. Our treatment will be merely formal and we shall not explain why differential 2-forms are ways of measuring two-dimensional areas. We see a differential 1-form in Ω as a formal combination $P(p,q)dp + Q(p,q)dq$ where P and Q are smooth real-valued functions defined in Ω. For instance, the differentials dp^* and dq^* of the components of the transformation ψ considered above are differential 1-forms

$$dp^* = \frac{\partial p^*}{\partial p}dp + \frac{\partial p^*}{\partial q}dq, \quad dq^* = \frac{\partial q^*}{\partial p}dp + \frac{\partial q^*}{\partial q}dq.$$

Two differential 1-forms ω and ω' give rise, via the *exterior* or wedge product $\wedge$, to a new entity $\omega \wedge \omega'$ called a differential 2-form. The exterior product is bilinear, so that, for instance,

$$\begin{aligned} dp^* \wedge dq^* &= \frac{\partial p^*}{\partial p}\frac{\partial q^*}{\partial p}dp \wedge dp + \frac{\partial p^*}{\partial p}\frac{\partial q^*}{\partial q}dp \wedge dq \\ &+ \frac{\partial p^*}{\partial q}\frac{\partial q^*}{\partial p}dq \wedge dp + \frac{\partial p^*}{\partial q}\frac{\partial q^*}{\partial q}dq \wedge dq. \end{aligned}$$

The exterior product is skew-symmetric. In particular, it holds that

$$dp \wedge dp = dq \wedge dq = 0, \quad dp \wedge dq = -dq \wedge dp.$$

Thus

$$dp^* \wedge dq^* = \left(\frac{\partial p^*}{\partial p}\frac{\partial q^*}{\partial q} - \frac{\partial p^*}{\partial q}\frac{\partial q^*}{\partial p}\right) dp \wedge dq$$

and from (2.5) we see that conservation of area is equivalent to

$$dp^* \wedge dq^* = dp \wedge dq.$$

This usually provides a convenient way of checking preservation of area.

2.5 Symplectic transformations

It is now time to consider the case $d > 1$. Is there something analogous to area that is being conserved by Hamiltonian solution operators? The $2d$-dimensional volume in Ω appears to be a natural candidate and indeed this volume *is* conserved. However this is not what we really want. What does the trick is to consider oriented *two-dimensional* surfaces D in Ω, to find the projections D_i, $i = 1, \ldots, d$ onto the d two-dimensional planes of the variables (p_i, q_i) and sum the two-dimensional oriented areas of these projections. This yields a number $m(D)$. It can be proved (see e.g. Arnold (1989) Section 44) that the solution operator of (1.1) preserves m: $m(\Phi_H(t, t_0)(D)) = m(D)$ whenever D is contained in the domain of $\Phi_H(t, t_0)$. Now transformations that have this preservation property are called *symplectic* or *canonical*, so that we have the theorem:

Theorem 2.1 *For each t, t_0, the solution operator $\Phi_H(t, t_0)$ of a Hamiltonian system (1.1) is a symplectic (or canonical) transformation.*

Furthermore, if Ω is simply connected (i.e., each closed curved in Ω may be shrunk down to a single point without leaving Ω), then the converse is also true: an m-preserving differential system is a Hamiltonian system, see Arnold (1989), Section 40D (once more, if Ω fails to be simply connected then preservation of m implies that the system is locally Hamiltonian). In this respect the symplecticness of the solution operator is the hallmark of Hamiltonian systems and once more the dynamical features that are specific to Hamiltonian problems can be traced back to the symplectic character of the solution operator.

In the classical literature the preservation of the sum $m(D)$ of oriented areas was referred to as conservation of the *Poincaré integral invariant*.

The condition (2.6) that we used to decide whether a transformation ψ in the plane was area-preserving or otherwise is generalized to read

$$\forall (\mathbf{p}, \mathbf{q}) \in \Omega, \quad \psi'^T J \psi' = \frac{\partial(\mathbf{p}^*, \mathbf{q}^*)}{\partial(\mathbf{p}, \mathbf{q})}^T J \frac{\partial(\mathbf{p}^*, \mathbf{q}^*)}{\partial(\mathbf{p}, \mathbf{q})} = J. \quad (2.7)$$

Note that the matrix J has the property that, for each pair $(\mathbf{v}, \mathbf{w})$ of vectors in $\mathcal{R}^{2d}$, $\mathbf{v}^T J \mathbf{w}$ represents the sum of the oriented two-dimensional areas of the d parallelograms that result from projecting the parallelogram determined by $\mathbf{v}$ and $\mathbf{w}$ onto the planes of the variables (p_i, q_i).

Differential forms can also be used. In the present context, 1-differential forms are formal expressions of the form $P_1 dp_1 + \cdots + P_d dp_d + Q_1 dq_1 + \cdots + Q_d dq_d$, with P_i and Q_i smooth real-valued functions defined in Ω. Again, two 1-forms give rise to a 2-form via the exterior product. The transformation ψ is symplectic if and

only if
$$dp_1^* \wedge dq_1^* + \cdots + dp_d^* \wedge dq_d^* = dp_1 \wedge dq_1 + \cdots + dp_d \wedge dq_d,$$
a relation that we more compactly rewrite as
$$d\mathbf{p}^* \wedge d\mathbf{q}^* = d\mathbf{p} \wedge d\mathbf{q}.$$

2.6 Conservation of volume

Let $\Phi_H(t, t_0)$ play the role of ψ in (2.7) and take determinants. The result is that $\det(\Phi_H(t, t_0)')$ is either $+1$ or -1. The value -1 is excluded since, by Liouville's theorem, the solution operator of any differential system has a Jacobian matrix with positive determinant. Hence $\det(\Phi_H(t, t_0)') \equiv 1$: Hamiltonian systems preserve the oriented volume in $\mathcal{R}^{2d}$ or, in other words, points in phase space convected by the solutions of Hamiltonian systems behave like particles of an incompressible fluid flow. Note that preservation of volume $\det(\psi') \equiv 1$ is a direct generalization to $d > 1$ of the property (2.5). However, when going from $d = 1$ to $d > 1$, the right generalization of preservation of area is symplecticness rather than preservation of volume. Symplecticness *characterizes* Hamiltonian solution operators; conservation of volume is a much weaker property shared by some non-Hamiltonian systems.

Conservation of volume is the key to the phenomenon of Poincaré *recurrence:* if, for all values of t, a bounded subdomain Σ of the phase space Ω is mapped into itself by a Hamiltonian flow $\phi_{t,H}$ then, as t increases, points in Ω moved by $\phi_{t,H}$ return repeatedly to the neighbourhood of their initial position (Arnold (1989), Section 16).

CHAPTER 3

Numerical methods

3.1 Numerical integrators

We now review some basic facts of the theory of numerical methods for the integration of systems of differential equations. For the time being, we restrict our attention to *one-step* methods: multistep methods are only considered in Chapter 14. General references on numerical integrators are Butcher (1987), Haireret al. (1987), Hairer and Wanner (1991), Lambert (1991).

Even though we are mainly interested in Hamiltonian problems, it is useful to present the methods as applied to general systems of differential equations

$$\frac{d\mathbf{y}}{dt} = \mathbf{F}(\mathbf{y}, t), \qquad (3.1)$$

where $\mathbf{F}$ represents a smooth $\mathcal{R}^D$-valued mapping defined in $\Omega \times I$, with Ω a domain in $\mathcal{R}^D$ and I an interval of the real line. Of course the Hamiltonian system (1.1) is just a particular instance of (3.1) with $D = 2d$ and $\mathbf{F} = J^{-1}\nabla H$ (cf. (1.2)). The solution operator (cf. Subsection 2.1) for (3.1) is denoted by $\Phi_\mathbf{F}(t, t_0)$. If the system is *autonomous*, $\mathbf{F} = \mathbf{F}(\mathbf{y})$, then we write $\phi_{t,\mathbf{F}}$ for the corresponding t-flow. If $\mathbf{y}^n$ denotes the numerical approximation at time t_n to the value $\mathbf{y}(t_n)$ of a solution of (3.1), then a (one-step) numerical method generates a smooth mapping $\Psi_\mathbf{F}(t_{n+1}, t_n)$ that effects the transition *(step)* from one *time level* to the next

$$\mathbf{y}^{n+1} = \Psi_\mathbf{F}(t_{n+1}, t_n)\mathbf{y}^n. \qquad (3.2)$$

The quantity $h_{n+1} = t_{n+1} - t_n$ is called the *step size* or *step length*. The simplest numerical method is the well-known (explicit) Euler rule,

$$\mathbf{y}^{n+1} = \mathbf{y}^n + (t_{n+1} - t_n)\mathbf{F}(\mathbf{y}^n, t_n), \qquad (3.3)$$

for which

$$\Psi_\mathbf{F}(t_{n+1}, t_n)\mathbf{y} = \mathbf{y} + (t_{n+1} - t_n)\mathbf{F}(\mathbf{y}, t_n).$$

Some important classes of methods are presented later in this chapter.

The domain of $\Psi_{\mathbf{F}}(t_{n+1}, t_n)$ need not be, for each t_{n+1} and t_n in I, the whole Ω. In fact for *implicit* methods, where the actual computation of $\mathbf{y}^{n+1}$ involves the solution of some system of equations, it is often the case that, for given $\mathbf{y}^n$ and t_n, the new approximation $\Psi_{\mathbf{F}}(t_{n+1}, t_n)\mathbf{y}^n$ is only defined if $t_{n+1} - t_n$ is sufficiently small.

For all methods of practical interest, if $\mathbf{F}$ happens to be independent of t, i.e., if (3.1) is autonomous, then $\Psi_{\mathbf{F}}(t_{n+1}, t_n)$ depends only on t_{n+1}, t_n through their difference h_{n+1}. In this case, in analogy with our notation for flows, we write $\psi_{h_{n+1}, \mathbf{F}}$ rather than $\Psi_{\mathbf{F}}(t_{n+1}, t_n)$.

If (3.1) happens to be the Hamiltonian system $\mathcal{S}_H$ with Hamiltonian H, then we write $\Psi_H(t_{n+1}, t_n)$ rather than $\Psi_{\mathbf{F}}(t_{n+1}, t_n)$ (or $\psi_{h_{n+1}, H}$ rather than $\psi_{h_{n+1}, \mathbf{F}}$ in the autonomous case).

Clearly for the method (3.2) to make sense it is necessary that $\Psi_{\mathbf{F}}(t_{n+1}, t_n)\mathbf{y}$ approximates the true $\Phi_{\mathbf{F}}(t_{n+1}, t_n)\mathbf{y}$. The difference

$$\Phi_{\mathbf{F}}(t_{n+1}, t_n)\mathbf{y} - \Psi_{\mathbf{F}}(t_{n+1}, t_n)\mathbf{y} \qquad (3.4)$$

is called the *local error* (at $\mathbf{y}$). For instance, for Euler's method the local error is given by

$$\bar{\mathbf{y}}(t_{n+1}) - \mathbf{y} - (t_{n+1} - t_n)\mathbf{F}(\mathbf{y}, t)$$
$$= \bar{\mathbf{y}}(t_{n+1}) - \bar{\mathbf{y}}(t_n) - (t_{n+1} - t_n)\dot{\bar{\mathbf{y}}}(t_n), \qquad (3.5)$$

where $\bar{\mathbf{y}}(t)$ is the solution of (3.1) that at time t_n takes the value $\mathbf{y}$.

A method of the form (3.2) is of *order (of consistency) r*, r an integer, if, as $t_{n+1} \to t_n$, the local error is $O(h_{n+1}^{r+1})$ whenever $\mathbf{F}$ is suitably smooth. In the case of Euler's method, Taylor expansion of (3.5) shows that the local error equals

$$\frac{h_{n+1}^2}{2} \frac{d^2 \bar{y}}{dt^2}(t_n) + O(h_{n+1}^3), \qquad (3.6)$$

so that the method is of order 1. *Consistency* means of course order ≥ 1.

Given an initial condition $\mathbf{y}^0$ at time t_0 and a grid $t_0 < t_1 < \ldots < t_N$, the numerical approximation at time t_N is found by iterating the mapping Ψ, i.e.

$$\mathbf{y}^N = \Psi_{\mathbf{F}}(t_N, t_{N-1}) \ldots \Psi_{\mathbf{F}}(t_1, t_0)\, \mathbf{y}^0.$$

This should be compared with the situation for the true solution,

STIFF PROBLEMS   27

for which
$$\mathbf{y}(t_N) = \Phi_\mathbf{F}(t_N, t_{N-1}) \ldots \Phi_\mathbf{F}(t_1, t_0)\, \mathbf{y}^0 = \Phi_\mathbf{F}(t_N, t_0)\, \mathbf{y}^0.$$
Here we have used the operator equation satisfied by the solution operator, cf. (2.2).

For a method of order r, it holds that, for t_N ranging in a bounded interval, the *global errors* $\mathbf{y}^N - \mathbf{y}(t_N)$ are $O(h^r)$, $h \to 0$, with h the maximum step size $\max_n h_{n+1}$ (Hairer et al. (1987), Section II.3). The exponent of h is now r, while local errors behave as $O(h^{r+1})$; this is of course related to the fact that as the step sizes are reduced it is necessary to take more steps to reach a given location in the t-axis.

3.2 Stiff problems

Stiff problems are a very important class of problems of the form (3.1). Let us present a particular instance.

Example 3.1 Assume that the Euler rule (3.3) is applied with constant step size h to the scalar problem
$$\dot{y} = \lambda y, \tag{3.7}$$
where $\lambda < 0$ is a constant. The computed points satisfy
$$y^{n+1} = (1 + h\lambda) y^n$$
and hence *grow* exponentially with n unless
$$|1 + h\lambda| \leq 1. \tag{3.8}$$
Since the true solution itself decays with t, the method produces very large (global) errors unless (3.8) is satisfied. If λ is very large in magnitude, (3.8) is a severe restriction on h. However, it is less severe than the restriction on h deriving from the requirement that the local error should be reasonably small, i.e., the requirement that h should be in line with the time scale in which the solution itself varies. For instance, it is easy to show using (3.6) that we would roughly require $h \leq 0.3/|\lambda|$ to have local errors of about 5%. Therefore the existence of a so-called *absolute stability restriction* $h \leq 2/|\lambda|$ is not a serious drawback of the method, as applied to (3.7). The choice of step size is dictated by accuracy considerations. However let us now consider the slightly more complicated nonhomogeneous *stiff* problem
$$\dot{y} = \lambda y + g(t), \ \lambda = -10^6, \ g(t) = \cos t - \lambda \sin t, \ y(0) = 1, \tag{3.9}$$

with solution
$$y(t) = \sin t + \exp(-10^6 t). \qquad (3.10)$$
After a short transient, (3.10) is virtually identical to the $\sin t$ function, and steps of length, say, $h = 0.1$ would be reasonable to keep the local error small. Nevertheless, from the definition of local error it follows easily that
$$y(t_{n+1}) - y^{n+1} = (1 - 10^6 h)[y(t_n) - y^n] + T_n$$
where T_n denotes the local error at $y(t_n)$. Accordingly, the global errors $y(t_n) - y^n$ will grow exponentially with n unless h satisfies the absolute stability restriction $h \leq 2/|\lambda| = 2 \times 10^{-6}$. This renders Euler's rule unsuitable for (3.9). □

The previous example shows some of the features typical of stiff problems. One of them is the existence of several time scales; in (3.9) these are the (slow) time scale of variation of the $\sin t$ function and the (fast) relaxation scale in the term $\exp(-10^6 t)$. Another feature is the large size of the *Lipschitz constant* L of $\mathbf{F}$ defined by
$$\|\mathbf{F}(\mathbf{u},t) - \mathbf{F}(\mathbf{v},t)\| \leq L \|\mathbf{u} - \mathbf{v}\|$$
for all $\mathbf{u}$ and $\mathbf{v}$ in Ω and all t in I. In the example this constant equals $|\lambda| = 10^6$. Due to severe stability limits on the step size, stiff problems cannot be efficiently integrated by explicit methods (Hairer and Wanner (1991), Dekker and Verwer (1984)). Unfortunately, stiff problems occur frequently in many areas of application including the time integration of evolutionary partial differential equations, see e.g. Sanz-Serna and Verwer (1989). In these applications implicit methods having good stability properties such as A- or B-stability should be used.

3.3 Runge-Kutta methods

3.3.1 The class of Runge-Kutta methods

Let us now present some important families of numerical methods. A *Runge-Kutta* (RK) method with s *stages* is specified by a *tableau* of real constants

$$\begin{array}{c|ccc} & a_{11} & \cdots & a_{1s} \\ & \vdots & \ddots & \vdots \\ & a_{s1} & \cdots & a_{ss} \\ \hline & b_1 & \cdots & b_s \end{array} \qquad (3.11)$$

The numbers b_i are the *weights* of the method.

When applied to the system (3.1), the method (3.11) advances the numerical solution from time t_n to time $t_{n+1} = t_n + h_{n+1}$ through the relation

$$\begin{aligned} \mathbf{y}^{n+1} &= \Psi_\mathbf{F}(t_{n+1}, t_n)\mathbf{y}^n \\ &= \mathbf{y}^n + h_{n+1} \sum_{i=1}^{s} b_i \mathbf{F}(\mathbf{Y}_i, t_n + c_i h_{n+1}), \end{aligned} \quad (3.12)$$

where

$$c_i = \sum_{j=1}^{s} a_{ij}, \quad i = 1, \ldots, s \quad (3.13)$$

are the *abscissae* of the method and the vectors $\mathbf{Y}_i$ are the so-called internal *stages*. The $\mathbf{Y}_i$ depend of course on n, but this dependence is not reflected in the notation. The internal stages are determined by the relations

$$\mathbf{Y}_i = \mathbf{y}^n + h_{n+1} \sum_{j=1}^{s} a_{ij} \mathbf{F}(\mathbf{Y}_j, t_n + c_j h_{n+1}), \quad i = 1, \ldots, s. \quad (3.14)$$

If $a_{ij} = 0$ whenever $i \leq j$, the equations (3.14) provide a recursion for explicitly computing each $\mathbf{Y}_i$ in terms of the preceding stages:

$$\begin{aligned} \mathbf{Y}_1 &= \mathbf{y}^n, \\ \mathbf{Y}_2 &= \mathbf{y}^n + h_{n+1} a_{21} \mathbf{F}(\mathbf{Y}_1, t_n + c_1 h_{n+1}), \\ \mathbf{Y}_3 &= \mathbf{y}^n + h_{n+1} a_{31} \mathbf{F}(\mathbf{Y}_1, t_n + c_1 h_{n+1}) \\ &\quad + h_{n+1} a_{32} \mathbf{F}(\mathbf{Y}_2, t_n + c_2 h_{n+1}), \end{aligned}$$

$\ldots \ldots \ldots \ldots \ldots \ldots \ldots \ldots \ldots \ldots \ldots \ldots \ldots$

The method is then called *explicit*. The computation of one step with an explicit RK method thus requires s evaluations of the function $\mathbf{F}$. Explicit RK methods are of course widely used algorithms in numerical differential equations.

When the method is *implicit*, (3.14) provides a coupled system of $s \times D$ algebraic equations for the $s \times D$ components of the stage vectors. The computational cost per step of an implicit RK method is then definitely higher than that of an explicit RK method and therefore implicit formulae have only been used for stiff problems, where their better stability properties make up for the computational cost.

A particular case of implicit methods, called *diagonally implicit* methods, occurs when $a_{ij} = 0$ for $i < j$. Then the solution of (3.14) requires the successive solution of s D-dimensional nonlinear systems

$$\begin{aligned} \mathbf{Y}_1 &= \mathbf{y}^n + h_{n+1} a_{11} \mathbf{F}(\mathbf{Y}_1, t_n + c_1 h_{n+1}), \to \mathbf{Y}_1, \\ \mathbf{Y}_2 &= \mathbf{y}^n + h_{n+1} a_{21} \mathbf{F}(\mathbf{Y}_1, t_n + c_1 h_{n+1}) \\ &\quad + h_{n+1} a_{22} \mathbf{F}(\mathbf{Y}_2, t_n + c_2 h_{n+1}), \to \mathbf{Y}_2, \\ &\dots \dots \dots \end{aligned}$$

This is far less demanding than the solution of a single $s \times D$ system that would be required for a general implicit method.

3.3.2 Collocation methods. Gauss methods

Many useful implicit RK methods can be derived via the idea of collocation. One starts by choosing the number of stages s and distinct abscissae c_i, $i = 1, \dots, s$. Then, a polynomial $\mathbf{u}_n(t)$, of degree $\leq s$, is determined such that $\mathbf{u}_n(t_n) = \mathbf{y}_n$ and $\mathbf{u}_n(t)$ satisfies the differential equation at the points $t_n + c_i h_{n+1}$, $i = 1, \dots, s$,

$$\mathbf{u}'_n(t_n + c_i h_{n+1}) = \mathbf{F}(\mathbf{u}_n(t_n + c_i h_{n+1}), t_n + c_i h_{n+1}).$$

Finally one sets $\mathbf{y}^{n+1} = \mathbf{u}_n(t_{n+1})$. The overall procedure can be proved to define an RK scheme (Hairer et al. (1987), Chapter II, Theorem 7.6). The internal stage vectors $\mathbf{Y}_i$ turn out to be the values of the collocation polynomial at the intermediate points $t_n + c_i h_{n+1}$, i.e.

$$\mathbf{u}_n(t_n + c_i h_{n+1}) = \mathbf{Y}_i, \quad i = 1, \dots, s, \qquad (3.15)$$

so that in particular the collocation conditions may be rewritten as

$$\mathbf{u}'_n(t_n + c_i h_{n+1}) = \mathbf{F}(\mathbf{Y}_i, t_n + c_i h_{n+1}), \quad i = 1, \dots, s. \qquad (3.16)$$

The weights b_i of the RK scheme obtained by the collocation procedure coincide with the weights of the interpolatory quadrature rule in the interval $[0, 1]$ based on the abscissae c_i. This means that, for the given c_i, the b_i, $i = 1, \dots, s$, are the unique choice for which the quadrature formula

$$\int_0^1 \phi(t)\, dt \approx \sum_{i=1}^s b_i \phi(c_i) \qquad (3.17)$$

gives no error whenever the integrand ϕ is a polynomial of degree $\leq s - 1$.

The order of any s-stage RK formula obtained by collocation is $\geq s$. To achieve order $s+r$, $r = 1, 2, \dots$ it is necessary and sufficient that (3.17) integrates exactly polynomials of degree $\leq s + r - 1$.

RUNGE-KUTTA METHODS

Standard results on numerical quadrature show then that it is not possible to have $r > s$. Furthermore there is a unique choice of c_i for which $r = s$. The resulting RK-collocation method is called *Gauss* or *Kuntzmann-Butcher* s-stage method and has order $2s$. No other RK method (either obtainable by collocation or otherwise) achieves order $2s$ with s stages.

The first Gauss methods possess the following tableaux. When $s = 1$ we have

$$\begin{array}{c|c} \frac{1}{2} & \frac{1}{2} \\ \hline & 1 \end{array}.$$

Thus the method reads

$$\mathbf{y}^{n+1} = \mathbf{y}^n + h_{n+1}\mathbf{F}\left(\mathbf{Y}_1, t_n + \frac{1}{2}h_{n+1}\right), \qquad (3.18)$$

$$\mathbf{Y}_1 = \mathbf{y}^n + \frac{h_{n+1}}{2}\mathbf{F}\left(\mathbf{Y}_1, t_n + \frac{1}{2}h_{n+1}\right). \qquad (3.19)$$

Multiplication by 2 of the last equation and subtraction from the first reveals that

$$\mathbf{Y}_1 = \frac{1}{2}(\mathbf{y}^n + \mathbf{y}^{n+1}) \qquad (3.20)$$

and therefore (3.18) can be rewritten as

$$\mathbf{y}^{n+1} = \mathbf{y}^n + h_{n+1}\mathbf{F}\left(\frac{1}{2}(\mathbf{y}^n + \mathbf{y}^{n+1}); t_n + \frac{1}{2}h_{n+1}\right);$$

in other words we are dealing with the familiar *implicit midpoint rule*.

For implementation purposes it is often advisable to solve (3.19) for $\mathbf{Y}_1$ and then find $\mathbf{y}^{n+1}$ via the relation

$$\mathbf{y}^{n+1} = 2\mathbf{Y}_1 - \mathbf{y}^n$$

that follows from (3.20).

For $s = 2$ (order 4) the tableau of the Gauss method is

$$\begin{array}{c|cc} \frac{1}{4} & \frac{1}{4} & \frac{1}{4} - \frac{\sqrt{3}}{6} \\ \frac{1}{4} + \frac{\sqrt{3}}{6} & & \frac{1}{4} \\ \hline & \frac{1}{2} & \frac{1}{2} \end{array},$$

while for $s = 3$ (order 6) we have

$$\begin{array}{|ccc|}
\dfrac{5}{36} & \dfrac{2}{9} - \dfrac{\sqrt{15}}{15} & \dfrac{5}{36} - \dfrac{\sqrt{15}}{30} \\
\dfrac{5}{36} + \dfrac{\sqrt{15}}{24} & \dfrac{2}{9} & \dfrac{5}{36} - \dfrac{\sqrt{15}}{24} \\
\dfrac{5}{36} + \dfrac{\sqrt{15}}{30} & \dfrac{2}{9} + \dfrac{\sqrt{15}}{15} & \dfrac{5}{36} \\
\hline
\dfrac{5}{18} & \dfrac{4}{9} & \dfrac{5}{18}
\end{array}.$$

Finally, for $s = 4$ (order 8) the tableau is

$$\begin{array}{|cccc|}
\omega_1 & \omega_1' - \omega_3 + \omega_4' & \omega_1' - \omega_3 - \omega_4' & \omega_1 - \omega_5 \\
\omega_1 - \omega_3' + \omega_4 & \omega_1' & \omega_1' - \omega_5' & \omega_1 - \omega_3' - \omega_4 \\
\omega_1 + \omega_3' + \omega_4 & \omega_1' + \omega_5' & \omega_1' & \omega_1 + \omega_3' - \omega_4 \\
\omega_1 + \omega_5 & \omega_1' + \omega_3 + \omega_4' & \omega_1' + \omega_3 - \omega_4' & \omega_1 \\
\hline
2\omega_1 & 2\omega_1' & 2\omega_1' & 2\omega_1
\end{array},$$

where

$$\omega_1 = \frac{1}{8} - \frac{\sqrt{30}}{144},$$

$$\omega_1' = \frac{1}{8} + \frac{\sqrt{30}}{144},$$

$$\omega_2 = \frac{1}{2}\left[\frac{15 + 2\sqrt{30}}{35}\right]^{1/2},$$

$$\omega_2' = \frac{1}{2}\left[\frac{15 - 2\sqrt{30}}{35}\right]^{1/2},$$

$$\omega_3 = \omega_2\left[\frac{1}{6} + \frac{\sqrt{30}}{24}\right],$$

$$\omega_3' = \omega_2'\left[\frac{1}{6} - \frac{\sqrt{30}}{24}\right],$$

$$\omega_4 = \omega_2\left[\frac{1}{21} + \frac{5\sqrt{30}}{168}\right],$$

$$\omega_4' = \omega_2'\left[\frac{1}{21} - \frac{5\sqrt{30}}{168}\right],$$

$$\omega_5 = \omega_2 - 2\omega_3,$$

$$\omega_5' = \omega_2' - 2\omega_3'.$$

The coefficients of the method with five stages can be found in Butcher (1964).

3.3.3 Existence and uniqueness of solutions in implicit methods

Let us now study whether, in implicit RK methods, the system (3.14) that defines the stages possesses, at each step, a unique solution. The following particular instance is illuminating.

Example 3.2 Consider the backward (or implicit) Euler method

$$\mathbf{y}^{n+1} = \mathbf{y}^n + (t_{n+1} - t_n)\mathbf{F}(\mathbf{y}^{n+1}, t_{n+1}), \qquad (3.21)$$

applied to the Hamiltonian problem in Example 2.1. The equation that defines q^{n+1} is of the second degree and hence easily solved. One finds that if the step length is h, then

$$q^{n+1} = \frac{1 \pm \sqrt{1 - 4hq^n}}{2h}.$$

Once q^{n+1} has been found, the new p^{n+1} is obtained by solving a linear equation. There are two things to be noticed. First, solutions exist only if the discriminant of the second-degree equation is positive. This sets an upper limit for the step size $h \leq 1/(4q^n)$. (Recall from Example 2.1 that for the true h-flow to be defined at q^n one finds a milder restriction $h < 1/q^n$.) Second, when the solution exists it is *not* unique: there are two *branches* of solutions. The branch obtained by choosing the minus sign for the square root approximates the true flow. In particular along this branch q^{n+1} approaches q^n as h tends to 0. The plus sign branch is *spurious;* the corresponding solutions do not relate at all to the differential equation being solved. The spurious solution tends to infinity as h goes to 0. □

In the general case the situation is similar. Solutions typically exist if the step size is suitably small. Furthermore when they exist, there are typically spurious branches along with the branch that approximates the theoretical solution. Provided that the step size is small, the spurious solutions are very large and will not be found by the iterative procedure used to solve the equations. For material on existence and uniqueness of RK solutions see e.g. Dekker and Verwer (1984), Chapter 5 and Sanz-Serna and Griffiths (1991). For spurious solutions see Iserles (1990) and Hairer *et al.* (1990), Iserles *et al.* (1991), Iserles and Stuart (1992).

Remark 3.1 In the very particular case where **F** in (3.1) satisfies a *uniform* Lipschitz condition with respect to **y** (in practice, if **F**

has partial derivatives of the first order with respect to the components of $\mathbf{y}$ that are uniformly bounded in $\mathcal{R}^D$), then it is not difficult to show (Butcher (1987), Corollary 341B) that the solution exists and *is unique* provided that the step size is small with respect to the inverse of the Lipschitz constant. □

3.4 Partitioned Runge-Kutta methods

In the integration of systems of differential equations (3.1) it is perfectly possible to integrate some components of the unknown vector $\mathbf{y}$ with a numerical method and the remaining components with a different numerical method. For instance, one may wish to do so if the system includes both stiff and nonstiff components. Let us assume that in (3.1) we write $\mathbf{y} = (\mathbf{p}, \mathbf{q})$, where the point $\mathbf{p}$ possesses d_1 components and $\mathbf{q}$ contains $d_2 = D - d_1$ components. If we partition correspondingly the right-hand side function $\mathbf{F} = (\mathbf{f}, \mathbf{g})$, then (3.1) becomes

$$\frac{d\mathbf{p}}{dt} = \mathbf{f}(\mathbf{p}, \mathbf{q}, t), \qquad \frac{d\mathbf{q}}{dt} = \mathbf{g}(\mathbf{p}, \mathbf{q}, t). \qquad (3.22)$$

From now on we only consider the special case

$$\frac{d\mathbf{p}}{dt} = \mathbf{f}(\mathbf{q}, t), \qquad \frac{d\mathbf{q}}{dt} = \mathbf{g}(\mathbf{p}). \qquad (3.23)$$

It is this particular format that will be of interest to us in our study of separable Hamiltonian problems. Let us assume that the p components are integrated by an RK formula and the q components with a different RK formula. The overall scheme is called a *Partitioned Runge-Kutta* (PRK) scheme and is specified by two *tableaux*

$$\begin{array}{c|ccc} & a_{11} & \cdots & a_{1s} \\ & \vdots & \ddots & \vdots \\ & a_{s1} & \cdots & a_{ss} \\ \hline & b_1 & \cdots & b_s \end{array} \quad , \quad \begin{array}{c|ccc} & A_{11} & \cdots & A_{1s} \\ & \vdots & \ddots & \vdots \\ & A_{s1} & \cdots & A_{ss} \\ \hline & B_1 & \cdots & B_s \end{array} \quad . \qquad (3.24)$$

The application of (3.24) to the system (3.23) results in the relations (cf. (3.14), (3.12))

$$\mathbf{P}_i = \mathbf{p}^n + h_{n+1} \sum_{j=1}^{s} a_{ij}\, \mathbf{f}(\mathbf{Q}_j, t_n + C_j h_{n+1}), \qquad (3.25)$$

$$\mathbf{Q}_i = \mathbf{q}^n + h_{n+1}\sum_{j=1}^{s} A_{ij}\,\mathbf{g}(\mathbf{P}_j), \qquad (3.26)$$

$i = 1,\ldots,s$, and

$$\mathbf{p}^{n+1} = \mathbf{p}^n + h_{n+1}\sum_{i=1}^{s} b_i\,\mathbf{f}(\mathbf{Q}_i, t_n + C_i h_{n+1}), \qquad (3.27)$$

$$\mathbf{q}^{n+1} = \mathbf{q}^n + h_{n+1}\sum_{i=1}^{s} B_i\,\mathbf{g}(\mathbf{P}_i). \qquad (3.28)$$

Here $\mathbf{P}_i$ and $\mathbf{Q}_i$ are the *stages* for the p and q variables, and

$$C_i = \sum_j A_{ij}$$

are the abscissae of the method. Of course an RK method is a particular instance of (3.24) where both tableaux just happen to have the same entries.

Remark 3.2 Above, the number of stages s_p for the p variables equals the number of stages s_q for the q variables. It is possible to conceive methods where those numbers are different. However such methods can be rewritten in the format (3.24) by adding dummy stages. See Remark 8.3. □

If it holds that $a_{ij} = A_{ij} = 0$ for $i \leq j$ in (3.24), then the method is explicit. An interesting situation arises if $a_{ij} = 0$ for $i < j$ and $A_{ij} = 0$ for $i \leq j$. Then, even though the a_{ij} tableau looks 'diagonally implicit' the computation is effectively explicit, due to the special structure of (3.23). In fact, in this case (3.25)–(3.26) read

$$\begin{aligned}
\mathbf{Q}_1 &= \mathbf{q}^n, \\
\mathbf{P}_1 &= \mathbf{p}^n + h_{n+1}a_{11}\mathbf{f}(\mathbf{Q}_1, t_n + C_1 h_{n+1}), \\
\mathbf{Q}_2 &= \mathbf{q}^n + h_{n+1}A_{21}\mathbf{g}(\mathbf{P}_1), \\
&\ldots
\end{aligned}$$

A similar situation arises if $a_{ij} = 0$ for $i \leq j$ and $A_{ij} = 0$ for $i < j$, or, more generally, if $a_{ij} = A_{ij} = 0$ for $i < j$ and for each i one amongst a_{ii} and A_{ii} vanishes. In the true diagonally implicit case we have $a_{ij} = A_{ij} = 0$ for $i < j$; then (3.25)–(3.26) demand the successive solution of s D-dimensional systems for the components of $(\mathbf{P}_i, \mathbf{Q}_i)$, $i = 1,\ldots,s$.

Considerations similar to those in Subsection 3.3.3 clearly apply to the case of implicit PRK methods.

3.5 Runge-Kutta-Nyström methods

Systems of differential equations of the special form

$$\frac{d\mathbf{v}}{dt} = \mathbf{f}(\mathbf{q}, t), \qquad \frac{d\mathbf{q}}{dt} = \mathbf{v}, \qquad (3.29)$$

or, equivalently, second-order systems

$$\frac{d^2\mathbf{q}}{dt^2} = \mathbf{f}(\mathbf{q}, t), \qquad (3.30)$$

can be efficiently integrated by means of Runge-Kutta-Nyström (RKN) methods (see e.g. Hairer *et al.* (1987), Section II.13). For the RKN procedure with *tableau*

$$\begin{array}{c|ccc} \gamma_1 & \alpha_{11} & \cdots & \alpha_{1s} \\ \vdots & \vdots & \ddots & \vdots \\ \gamma_s & \alpha_{s1} & \cdots & \alpha_{ss} \\ \hline & \beta_1 & \cdots & \beta_s \\ \hline & b_1 & \cdots & b_s \end{array} \qquad (3.31)$$

the intermediate *stages* $\mathbf{Q}_i$ are defined by

$$\mathbf{Q}_i = \mathbf{q}^n + h_{n+1}\gamma_i \mathbf{v}^n + h_{n+1}^2 \sum_{j=1}^{s} \alpha_{ij}\mathbf{f}(\mathbf{Q}_j, t_n + \gamma_j h_{n+1}), \qquad (3.32)$$

and the approximations at the next time level are

$$\mathbf{v}^{n+1} = \mathbf{v}^n + h_{n+1}\sum_{i=1}^{s} b_i \mathbf{f}(\mathbf{Q}_i, t_n + \gamma_i h_{n+1}),$$

$$\mathbf{q}^{n+1} = \mathbf{q}^n + h_{n+1}\mathbf{v}^n + h_{n+1}^2 \sum_{i=1}^{s} \beta_i \mathbf{f}(\mathbf{Q}_i, t_n + \gamma_i h_{n+1}).$$

For explicit methods $\alpha_{ij} = 0$ for $i \geq j$.

If a consistent PRK method (or in particular a consistent RK method) is applied to (3.29) then, after eliminating the stages $\mathbf{V}_i$ corresponding to the $\mathbf{v}$ variables, it is readily seen that the result is equivalent to the application of the RKN method with coefficients

$$\beta_i = \sum_{j=1}^{s} B_j a_{ji}, \quad i = 1, \ldots, s,$$

$$\gamma_i = \sum_{j=1}^{s} A_{ij}, \quad i = 1, \ldots, s, \qquad (3.33)$$

$$\alpha_{ij} = \sum_{k=1}^{s} A_{ik}a_{kj}, \quad i,j = 1,\ldots,s.$$

The *weights* b_i of the RKN method just coincide with the b_i of the PRK scheme. We say that the RKN method is *induced* by the PRK method.

Example 3.3 For the two-stage Gauss method, the induced RKN method has the tableau

$$
\begin{array}{c|cc}
\frac{1}{2} - \frac{\sqrt{3}}{6} & \frac{1}{24} & \frac{1}{8} - \frac{\sqrt{3}}{12} \\
\frac{1}{2} + \frac{\sqrt{3}}{6} & \frac{1}{8} + \frac{\sqrt{3}}{12} & \frac{1}{24} \\
\hline
& \frac{1}{4} + \frac{\sqrt{3}}{12} & \frac{1}{4} - \frac{\sqrt{3}}{12} \\
\hline
& \frac{1}{2} & \frac{1}{2}
\end{array}.
$$

This is easily checked by using (3.33). □

Considerations similar to those in Subsection 3.3.3 apply to implicit RKN formulae.

3.6 Composition of methods. Adjoints

3.6.1 Composing methods

We now leave our study of specific families of numerical algorithms and return to the study of the general class (3.2).

It is expedient to consider the issue of composition of methods, as this plays a role in later developments: it turns out that many symplectic methods are compositions of simpler methods. If $\Psi_{\mathrm{F}}^1(t_{n+1}, t_n)$ and $\Psi_{\mathrm{F}}^2(t_{n+1}, t_n)$ are numerical methods and θ is a real parameter, then the concatenation of a step of length θh_{n+1} with the first method and a step of length $(1-\theta)h_{n+1}$ with the second method, i.e., the mapping

$$\Psi_{\mathrm{F}}^2(t_{n+1}, \tau)\Psi_{\mathrm{F}}^1(\tau, t_n), \quad \tau = \theta t_{n+1} + (1-\theta)t_n,$$

defines a new numerical method $\Psi_{\mathrm{F}}(t_{n+1}, t_n)$ referred to as a *composition* of methods Ψ^1 and Ψ^2.

Example 3.4 If Ψ^1 and Ψ^2 are s-stage RK methods with coefficients a_{ij}^1, b_i^1 and a_{ij}^2, b_i^2, then the composition is an RK procedure with $2s$ stages given by

$$\left|\begin{array}{cccccc} \theta a_{11}^1 & \cdots & \theta a_{1s}^1 & 0 & \cdots & 0 \\ \vdots & \ddots & \vdots & \vdots & \ddots & \vdots \\ \theta a_{s1}^1 & \cdots & \theta a_{ss}^1 & 0 & \cdots & 0 \\ \theta b_1^1 & \cdots & \theta b_s^1 & (1-\theta)a_{11}^2 & \cdots & (1-\theta)a_{1s}^2 \\ \vdots & \ddots & \vdots & \vdots & \ddots & \vdots \\ \theta b_1^1 & \cdots & \theta b_s^1 & (1-\theta)a_{s1}^2 & \cdots & (1-\theta)a_{ss}^2 \\ \hline \theta b_1^1 & \cdots & \theta b_s^1 & (1-\theta)b_1^2 & \cdots & (1-\theta)b_s^2 \end{array}\right. \quad (3.34)$$

If Ψ^1 has s_1 stages and Ψ^2 s_2 stages then the composition has $s_1 + s_2$ stages and obvious changes are needed in (3.34). Similar constructions apply in the PRK and RKN cases. □

3.6.2 Adjoints

In connection with the idea of composition of methods we consider the notion of the *adjoint* of a given method Ψ. By definition, this is the method $\bar{\Psi}$ such that

$$\bar{\Psi}_{\mathbf{F}}(t_n, t_{n+1})\Psi_{\mathbf{F}}(t_{n+1}, t_n)$$

is the identity mapping: stepping forward with the given method is just stepping backward with its adjoint. The adjoint of the adjoint is the given method. A method that is its own adjoint is called *symmetric*.

The adjoint of the (explicit) (forward) Euler method (3.3) is the implicit or backward Euler method (3.21). In fact, if $\mathbf{u} \to \mathbf{v}$ is a forward Euler step from t_n, then

$$\mathbf{u} = \mathbf{v} + h_{n+1}\mathbf{F}(\mathbf{v}, t_n),$$

an equality that we rearrange as

$$\mathbf{v} = \mathbf{u} + (-h_{n+1})\mathbf{F}(\mathbf{v}, t_{n+1} + (-h_{n+1})).$$

This shows that $\mathbf{v}$ arises as a backward Euler step of length $-h_{n+1}$ taken from $\mathbf{u}$ at time t_{n+1}.

For a method (3.2) of order r, the adjoint has also order r. Furthermore, in Ψ and $\bar{\Psi}$ the leading $O(h^{r+1})$ terms in the expansion of the local error are equal, if r is even, or opposite, if r is odd (Hairer *et al.* (1987), Chapter II, Theorem 8.4). It then follows that the order of a symmetric method is necessarily even.

COMPOSITION OF METHODS. ADJOINTS

The adjoint of a composition is found by composing the adjoints of the factors in reversed order. This is just a simple implication of the general rule for finding the inverse of a composition of mappings. In particular it follows that the concatenation of a step of length $h/2$ of a method Ψ with a step of length $h/2$ taken with its adjoint results in a *symmetric* method.

3.6.3 Finding Runge-Kutta, Partitioned Runge-Kutta and Runge-Kutta-Nyström adjoints

The adjoint of the RK method (3.11) is the s-stage RK method with coefficients

$$\bar{a}_{ij} = b_{s+1-j} - a_{s+1-i,s+1-j}, \qquad \bar{b}_j = b_{s+1-j}. \qquad (3.35)$$

These formulae are easily found by manipulating the equations as we did in the Euler case.

Some readers may wish to check that, for $s = 1, \ldots, 4$, the s-stage Gauss method in Subsection 3.3.2 is its own adjoint. The same happens for all values of s, i.e., Gauss methods are symmetric (Hairer et al. (1987), Chapter II, Theorem 8.8).

The adjoint of the PRK method (3.24) is the PRK method with s stages given by (3.35) along with

$$\bar{A}_{ij} = B_{s+1-j} - A_{s+1-i,s+1-j}, \qquad \bar{B}_j = B_{s+1-j}.$$

For the RKN method (3.31) the adjoint is the s-stage RKN scheme (Okunbor and Skeel (1992a))

$$\begin{aligned}
\bar{\gamma}_i &= 1 - \gamma_{s+1-i}, \\
\bar{\alpha}_{ij} &= b_{s+1-j} - \beta_{s+1-j} - \gamma_{s+1-i}b_{s+1-j} + \alpha_{s+1-i,s+1-j}, \\
\bar{\beta}_j &= b_{s+1-j} - \beta_{s+1-j}, \\
\bar{b}_j &= b_{s+1-j}.
\end{aligned}$$

CHAPTER 4

Order conditions

4.1 The order in Runge-Kutta methods

In this chapter we continue our revision of some basic facts about standard numerical methods. We now turn our attention to *order conditions* and start by reviewing the conditions that must be imposed on the elements of the RK tableau (3.11) if the formula is to achieve a prescribed order (Butcher (1987), Theorem 307B, Hairer *et al.* (1987), Chapter II, Theorem 2.13). These conditions are written in terms of *rooted trees*. The left half of Fig. 4.1 depicts the rooted trees with ≤ 4 *vertices;* the *roots* have been highlighted by means of a cross. The vertices of a rooted tree that are adjacent to the root are called the *sons* of the root. The sons of the remaining vertices are defined in an obvious recursive way. For instance in Fig. 4.2, the vertex labelled n is the son of the vertex labelled m. A vertex is said to be the *father* of its sons.

For the method to have order of consistency r, each rooted tree $\rho\tau$ with *order* $\leq r$, i.e., with $\leq r$ vertices, gives rise to a condition

$$\Phi(\rho\tau) = 1/\gamma(\rho\tau). \qquad (4.1)$$

Here the *density* $\gamma(\rho\tau)$ is an integer associated with $\rho\tau$ and the *elementary weight* $\Phi(\rho\tau)$ is a polynomial in the method coefficients a_{ij}, b_i.

The recipe for writing down $\Phi(\rho\tau)$ for a given $\rho\tau$ is simple. We illustrate it for the graph in Fig. 4.2. We associate with each vertex a summation index $i, j, k, \ldots$ Then the corresponding Φ is given by the summation

$$\Phi = \sum_{i,j,k,l,m,n,o=1}^{s} b_i a_{ij} a_{jk} a_{il} a_{im} a_{mn} a_{no}.$$

The summation index associated with the root appears as a subscript for the letter b and, for a rooted tree of order r, there are $r-1$ coefficients a_{ij} where $[i,j]$ runs through all [father, son] pairs.

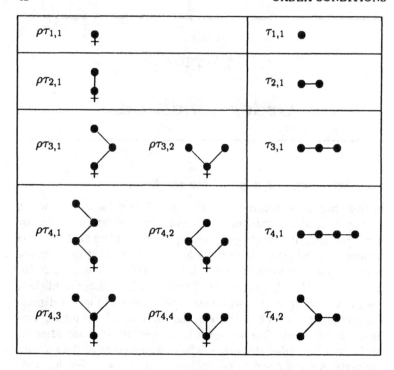

Figure 4.1. *Rooted r-trees and r-trees, r = 1,2,3,4*

The density γ is computed recursively. The density of the rooted tree with one vertex $\rho\tau_{1,1}$ (see Fig. 4.1) is, by definition, unity. The density of a rooted tree $\rho\tau$ of order r equals r times the product of the densities of the rooted trees that arise when the root of $\rho\tau$ is chopped off. For instance, in Fig. 4.1

$$\gamma(\rho\tau_{4,2}) = 4\gamma(\rho\tau_{2,1})\gamma(\rho\tau_{1,1})$$

and, in turn,

$$\gamma(\rho\tau_{2,1}) = 2\gamma(\rho\tau_{1,1})\gamma(\rho\tau_{1,1}),$$

so that $\gamma(\rho\tau_{4,2}) = 8$.

As an illustration, let us write the order conditions for $r = 1, 2, 3$. For consistency ($r \geq 1$) we require, in connection with $\rho\tau_{1,1}$,

$$\sum_{i=1}^{s} b_i = 1. \qquad (4.2)$$

THE LOCAL ERROR IN RUNGE-KUTTA METHODS

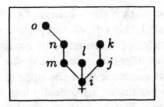

Figure 4.2. *A rooted tree of order 7*

For order $r \geq 2$ we further impose, in connection with $\rho\tau_{2,1}$,

$$\sum_{i,j=1}^{s} b_i a_{ij} = 1/2. \tag{4.3}$$

For order $r \geq 3$ we yet add, in view of $\rho\tau_{3,1}$,

$$\sum_{i,j,k=1}^{s} b_i a_{ij} a_{jk} = 1/6, \tag{4.4}$$

and, in view of $\rho\tau_{3,2}$,

$$\sum_{i,j,k=1}^{s} b_i a_{ij} a_{ik} = 1/3. \tag{4.5}$$

4.2 The local error in Runge-Kutta methods

It is expedient to study in some detail the local error (3.4) for the RK method (3.11). The analysis is simplified considerably if it is assumed that the system (3.1) is *autonomous*. There is no loss of generality in this assumption. In fact let us suppose that we are interested in integrating the *nonautonomous* system (3.1) with the initial condition $\mathbf{y}(t_0) = \mathbf{y}^0$. Let us construct a new (autonomous) initial value problem in $\mathcal{R}^{D+1}$ by setting

$$\frac{d\hat{\mathbf{y}}}{d\tau} = \hat{\mathbf{F}}(\hat{\mathbf{y}}), \quad \hat{\mathbf{y}}(\tau)\big|_{\tau=t_0} = (\mathbf{y}^0, t_0), \tag{4.6}$$

where the new dependent variable $\hat{\mathbf{y}}$ is given by $(\mathbf{y},t)$ and the new right-hand side function $\hat{\mathbf{F}}$ has components $(\mathbf{f},1)$. Note that $dt/d\tau = 1$ and the initial condition $t(\tau_0) = t_0$ imply that $t \equiv \tau$. It is then obvious that the first D components of the solution of the autonomous problem (4.6) provide the solution of the nonautonomous original problem. Conversely, by appending a $(D+1)$-th component equal to t to the (D-dimensional) solution of the original problem, we obtain the solution of (4.6). What is important to us here is that this equivalence between the autonomous and nonautonomous forms is preserved by consistent RK methods. In other words, if we integrate numerically the autonomous form and discard the last component of the approximation to $\hat{\mathbf{y}}$, then we obtain the same results as numerically integrating the original nonautonomous form. This is trivially checked. In practice it is the nonautonomous form that is integrated, but when it comes to *analysing* the local error or other properties of the method it is possible to assume that it is the autonomous form that has been dealt with.

Remark 4.1 Before proceeding any further, let us point out that if the original system is Hamiltonian, then the rewriting in autonomous form can be carried out so as to preserve the Hamiltonian form: t is seen as a new coordinate q_{d+1} and the corresponding conjugate momentum p_{d+1} is required to evolve according to $dp_{d+1}/d\tau = -\partial H/\partial t$. It is trivial to check that the variables $(p_1, \ldots, p_d, p_{d+1}, q_1, \ldots, q_{d+1})$ as functions of $\tau \equiv t \equiv q_{d+1}$ satisfy the Hamiltonian differential equations corresponding to the autonomous Hamiltonian function $H(p_1, \ldots, p_d, q_1, \ldots, q_{d+1}) + p_{d+1}$. Once more, consistent RK methods preserve the equivalence between the autonomous and nonautonomous forms. □

In view of the preceding considerations we assume in this section that the system (3.1) being integrated is autonomous. It can be shown (Hairer *et al.* (1987), Section II.3) that, for fixed $\mathbf{y}$, the Taylor expansion of $\psi_{h,\mathbf{F}}\mathbf{y}$ as a function of the step size h is given by

$$\psi_{h,\mathbf{F}}\mathbf{y} = \mathbf{y} + \sum_{m=1}^{\infty} \frac{h^m}{m!} \sum_{\rho\tau \in \mathcal{T}_m} \alpha(\rho\tau)\gamma(\rho\tau)\Phi(\rho\tau)\mathcal{F}(\rho\tau)(\mathbf{y}), \quad (4.7)$$

where $\mathcal{T}_m$ denotes the set of all rooted trees of order m, $\gamma(\rho\tau)$ and $\Phi(\rho\tau)$ are the density and weight we found above, $\alpha(\rho\tau)$ is an integer associated with $\rho\tau$ and $\mathcal{F}(\rho\tau)$ is the *elementary differential* of $\mathbf{F}$ corresponding to $\rho\tau$.

For a rooted tree $\rho\tau$ of order r, $\alpha(\rho\tau)$ is defined to be the number of its *monotonic labellings*, i.e., the number of ways in which the labels $\{1, 2, \ldots, r\}$ can be attached to the vertices of $\rho\tau$ in such a way that each vertex receives as a label an integer which is larger than the integer received by its father.

Example 4.1 For $\rho\tau_{4,2}$ in Fig. 4.1 $\alpha = 3$: the label 1 goes to the root, the label 3 may go to any of the remaining three vertices;

then the locations of labels 2, 4 are uniquely determined by the requirement that sons have higher labels than their fathers. □

Example 4.2 For all other rooted trees in Fig. 4.1 $\alpha = 1$. For instance in $\rho\tau_{3,2}$, the label 1 must go to the root and then 2 or 3 are the labels of the sons of the root. It does not matter whether, *in the figure*, the label 2 goes to the vertex to the right of the root or to the vertex to the left of the root. In either case, the father–son pairs are $(1,2)$ and $(1,3)$, so that both cases coincide from a combinatorial point of view: they are two ways of pictorially representing *the same* monotonic labelling. □

For each rooted tree $\rho\tau$, the elementary differential $\mathcal{F}(\rho\tau)$ is an $\mathcal{R}^D$-valued mapping defined in the domain Ω of $\mathbf{F}$. For the graph in Fig. 4.2, the i-th component of the vector $\mathcal{F}(\rho\tau)(\mathbf{y})$ is given by

$$\sum_{j,k,l,m,n,l,m,n,o=1}^{D} F^i_{j,l,m} F^j_k F^k F^l F^m_n F^n_o F^o,$$

where superscripts denote components and subscripts represent differentiation (thus $F^i_{j,l,m}$ is the partial derivative of order 3 of the i-th component of $\mathbf{F}$ with respect to the variables y_j, y_l, y_m). Partial derivatives are evaluated at $\mathbf{y}$. In general, there is an index associated with each vertex of the rooted tree and, for a rooted tree of order r, the terms being summed are products of r factors. Each factor is of the form $F^a_{b,c,\ldots}$ where $b, c, \ldots$ are the indices associated with the sons of the vertex with index a.

The true flow $\phi_{h,\mathbf{F}}$ has an expansion similar to (4.7), namely

$$\phi_{h,\mathbf{F}}\mathbf{y} = \mathbf{y} + \sum_{m=1}^{\infty} \frac{h^m}{m!} \sum_{\rho\tau \in T_m} \alpha(\rho\tau) \mathcal{F}(\rho\tau)(\mathbf{y}). \qquad (4.8)$$

The expansion of the local error is now obtained by subtracting (4.7) from (4.8). The meaning of the order conditions (4.1) becomes clear: when (4.1) holds for rooted trees of order $\leq r$, then the expansions of the method and of the true flow coincide up to the h^r terms and the local error is $O(h^{r+1})$.

4.3 The order in Partitioned Runge-Kutta methods

Let us now consider the case of a PRK method (3.24) applied to a partitioned system (3.23). Once again graph theory can be used to systematize the writing of the order conditions (Hairer *et al.* (1987), Section II.14). We now need *bicolour rooted trees* $\beta\rho\tau$, i.e., rooted trees with their vertices coloured either *white* or *black* in such a way that adjacent vertices receive distinct colours. Clearly each rooted tree gives rise to two bicolour rooted trees: the root can be coloured either black or white and the colour of the root recursively determines the colour of the remaining vertices. The left section of Fig. 4.3 depicts the bicolour rooted trees with four or fewer vertices. This should be compared with Fig. 4.1.

There is an order condition

$$\Phi(\beta\rho\tau) = 1/\gamma(\beta\rho\tau) \qquad (4.9)$$

for each $\beta\rho\tau$. The first of these are as follows. For $\beta\rho\tau_{1,1,w}$ and $\beta\rho\tau_{1,1,b}$ we respectively have (cf. (4.2))

$$\sum_{i=1}^{s} b_i = 1, \qquad \sum_{i=1}^{s} B_i = 1.$$

In connection with $\beta\rho\tau_{2,1,w}$ and $\beta\rho\tau_{2,1,b}$ we respectively write (cf. (4.3))

$$\sum_{i,j=1}^{s} b_i A_{ij} = 1/2, \qquad \sum_{i,j=1}^{s} B_i a_{ij} = 1/2. \qquad (4.10)$$

The general rule is that white vertices bring in lower case letters in the *elementary weight* Φ and black vertices bring in upper case letters. The *density* γ of a bicolour rooted tree is the same as the density of the underlying (uncoloured) rooted tree, e.g. $\gamma(\beta\rho\tau_{3,2,w}) = \gamma(\beta\rho\tau_{3,2,b}) = \gamma(\rho\tau_{3,2})$.

Remark 4.2 In the particular case in which the PRK reduces to an RK method, i.e., if both tableaux in (3.24) are the same, then it is clear that the order conditions of the method, when seen as a PRK scheme, reproduce, twice over, the order conditions of the method when seen as an RK scheme. □

4.4 The local error in Partitioned Runge-Kutta methods

Again it is convenient to assume that (3.23) is autonomous, i.e.

$$\frac{d\mathbf{p}}{dt} = \mathbf{f}(\mathbf{q}), \qquad \frac{d\mathbf{q}}{dt} = \mathbf{g}(\mathbf{p}). \qquad (4.11)$$

This implies no loss of generality because the nonautonomous case can be rewritten in autonomous form by considering t as a new q variable. (If (3.23) happens to be a Hamiltonian system, this rewriting is consistent with that discussed in Remark 4.1.)

THE LOCAL ERROR IN PARTITIONED RUNGE-KUTTA METHODS

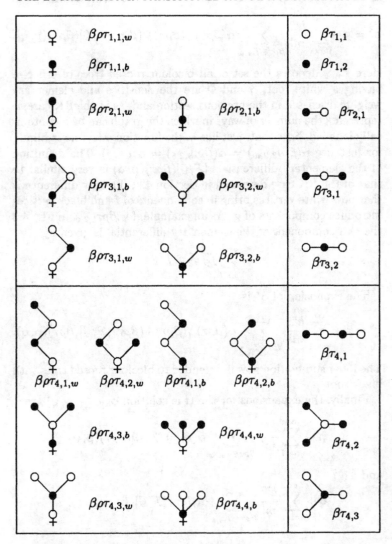

Figure 4.3. *Bicolour rooted r-trees and bicolour r-trees, $r=1,2,3,4$*

If we represent the PRK method as

$$(\mathbf{p}^*, \mathbf{q}^*) = \psi_{h,(\mathbf{f},\mathbf{g})}(\mathbf{p}, \mathbf{q}),$$

then the Taylor expansion of $\mathbf{p}^*$ as a function of the step size h reads

$$\mathbf{p}^* = \mathbf{p} + \sum_{m=1}^{\infty} \frac{h^m}{m!} \sum_{\beta\rho\tau \in T_{m,w}} \alpha(\beta\rho\tau)\, \gamma(\beta\rho\tau)\, \Phi(\beta\rho\tau)[\mathcal{F},\mathcal{G}](\beta\rho\tau)(\mathbf{p},\mathbf{q}).$$

Here $T_{m,w}$ denotes the set of all bicolour rooted trees of order m having a white root, γ and Φ are the densities and elementary weights discussed in the previous section and, as in the RK case, α represents the number of ways in which the graph can be monotonically labelled. Note that, just like γ, the function α is 'colour-blind', for instance $\alpha(\beta\rho\tau_{3,2,w}) = \alpha(\beta\rho\tau_{3,2,b}) = \alpha(\rho\tau_{3,2})$. The definition of the elementary differentials $[\mathcal{F},\mathcal{G}](\beta\rho\tau)(\mathbf{p},\mathbf{q})$ is very similar to that of the RK case discussed in Section 4.2; the only difference is that now white vertices bring in components of $\mathbf{f}$ and black vertices introduce components of $\mathbf{g}$. As an example, for $\beta\rho\tau_{4,2,w}$ in Fig. 4.3 the i-th component of the elementary differential is given by

$$\sum_{j,k,l} f^i_{jl}\, g^j_k\, g^l\, f^k.$$

The expansion of $\mathbf{q}^*$ is

$$\mathbf{q}^* = \mathbf{q} + \sum_{m=1}^{\infty} \frac{h^m}{m!} \sum_{\beta\rho\tau \in T_{m,b}} \alpha(\beta\rho\tau)\, \gamma(\beta\rho\tau)\, \Phi(\beta\rho\tau)\, [\mathcal{F},\mathcal{G}](\beta\rho\tau)(\mathbf{p},\mathbf{q}).$$

The inner summation now is extended to bicolour rooted trees with black root.

Finally, the expansions for the true solution $\phi_{h,(\mathbf{f},\mathbf{g})}(\mathbf{p},\mathbf{q})$ are

$$\mathbf{p} + \sum_{m=1}^{\infty} \frac{h^m}{m!} \sum_{\beta\rho\tau \in T_{m,w}} \alpha(\beta\rho\tau)\, [\mathcal{F},\mathcal{G}](\beta\rho\tau)(\mathbf{p},\mathbf{q})$$

and

$$\mathbf{q} + \sum_{m=1}^{\infty} \frac{h^m}{m!} \sum_{\beta\rho\tau \in T_{m,b}} \alpha(\beta\rho\tau)\, [\mathcal{F},\mathcal{G}](\beta\rho\tau)(\mathbf{p},\mathbf{q}).$$

4.5 The order in Runge-Kutta-Nyström methods

In the study of the order conditions for RKN methods (3.31), the relevant graphs are *special Nyström rooted trees* $\sigma\nu\rho\tau$. These graphs have two kinds of vertices, *fat* and *meagre*. The root is fat. Fat vertices have only meagre sons. A meagre vertex has, at most, one son, which is fat. The left section of Fig. 4.4 displays the special Nyström rooted trees with four or fewer vertices.

Let us assume, until further notice, that the coefficients in the

THE ORDER IN RUNGE-KUTTA-NYSTRÖM METHODS

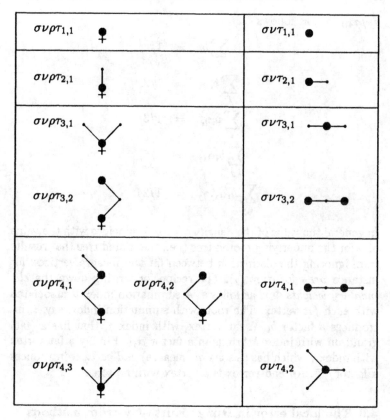

Figure 4.4. *Special Nyström rooted r-trees and special Nyström r-trees,* $r=1,2,3,4$

RKN tableau (3.31) satisfy

$$\beta_i = b_i(1 - \gamma_i), \quad i = 1, \ldots, s. \tag{4.12}$$

This is a so-called *simplifying assumption* that substantially reduces the number of order conditions to be considered, see Hairer et al. (1987) Chapter II, Lemma 13.13. Most methods of interest satisfy (4.12).

When *(4.12)* holds, there is an order condition

$$\Phi(\sigma\nu\rho\tau) = 1/\gamma(\sigma\nu\rho\tau) \tag{4.13}$$

for each $\sigma\nu\rho\tau$. The first of these, corresponding respectively to the special Nyström rooted trees $\sigma\nu\rho\tau_{1,1}$, $\sigma\nu\rho\tau_{2,1}$, $\sigma\nu\rho\tau_{3,1}$, $\sigma\nu\rho\tau_{3,2}$,

$\sigma\nu\rho\tau_{4,1}$, are as follows:

$$\sum_i b_i = 1,$$

$$\sum_i b_i \gamma_i = 1/2,$$

$$\sum_i b_i \gamma_i^2 = 1/3,$$

$$\sum_{i,j} b_i \alpha_{ij} = 1/6,$$

$$\sum_{i,j} b_i \alpha_{ij} \gamma_j = 1/24.$$

In general the value of the *density* $\gamma(\sigma\nu\rho\tau)$ coincides with the value of γ for the underlying rooted tree (i.e., the rooted tree that results from ignoring the distinction between fat and meagre vertices; for instance $\rho\tau_{4,2}$ for $\sigma\nu\rho\tau_{4,2}$). The recipe for writing down the *elementary weights* Φ is as follows. A summation index is associated with each *fat* vertex. The root, with summation index, say, i, introduces a factor b_i. A fat vertex, with index j, that has a (fat) grandson with index k brings in a factor α_{jk}. Finally a fat vertex with index j which has as sons m (meagre) *end-vertices* introduces a factor γ_j^m. An end-vertex is a vertex with no sons.

4.6 The local error in Runge-Kutta-Nyström methods

As in the RK and PRK cases, the study of the local error is best conducted under the assumption that the problem being integrated is autonomous. If we represent the method in the form

$$(\mathbf{v}^*, \mathbf{q}^*) = \psi_{h,\mathbf{f}}(\mathbf{v}, \mathbf{q}),$$

then the Taylor expansion of $\mathbf{v}^*$ is

$$\mathbf{v}^* = \mathbf{v} + \sum_{m=1}^{\infty} \frac{h^m}{m!} \sum_{\sigma\nu\rho\tau\in\mathcal{N}T_m} \alpha(\sigma\nu\rho\tau)\,\gamma(\sigma\nu\rho\tau)\,\Phi(\sigma\nu\rho\tau)\,f(\sigma\nu\rho\tau)(\mathbf{v},\mathbf{q});$$

the inner sum is extended to the set $\mathcal{N}T_m$ of all special Nyström rooted trees of order m. The symbols γ and Φ have been introduced above; α denotes once more the number of monotonic labellings (which obviously coincides with that of the underlying rooted tree);

$f(\sigma\nu\rho\tau)(\mathbf{v},\mathbf{q})$ denotes the elementary differential. The i-th component of the elementary differentials corresponding to $\sigma\nu\rho\tau_{4,1}$ and $\sigma\nu\rho\tau_{4,2}$ are respectively equal to

$$\sum_{jk} f_j^i f_k^j v_k, \quad \text{and} \quad \sum_{jk} f_{jk}^i f^j v_k.$$

There is an index per fat vertex or meagre end-vertex. A fat vertex with index j with (fat) grandsons with indices $k, l, \ldots$ and (meagre) end-vertex sons with indices $a, b, \ldots$ introduces a factor $f_{k,l,\ldots,a,b,\ldots}^j$. A meagre end-vertex son with index a introduces a factor v_a (subscripts in $\mathbf{v}$ denote components).

For the true solution the corresponding expansion is

$$\mathbf{v} + \sum_{m=1}^{\infty} \frac{h^m}{m!} \sum_{\sigma\nu\rho\tau \in \mathcal{N}T_m} \alpha(\sigma\nu\rho\tau) \, f(\sigma\nu\rho\tau)(\mathbf{v},\mathbf{q}).$$

Comparison of the two expansions above clarifies the meaning of (4.13).

For the numerical $\mathbf{q}^*$ the expansion is found to be

$$\mathbf{q}^* = \mathbf{q} + h\mathbf{v} + \sum_{m=2}^{\infty} \frac{h^m}{m!}$$
$$\sum_{\sigma\nu\rho\tau \in \mathcal{N}T_{m-1}} m\, \alpha(\sigma\nu\rho\tau)\, \gamma(\sigma\nu\rho\tau)\, \tilde{\Phi}(\sigma\nu\rho\tau)\, f(\sigma\nu\rho\tau)(\mathbf{v},\mathbf{q}). \quad (4.14)$$

Note that the inner summation now extends to trees of order $m-1$ rather than m. The tilde in $\tilde{\Phi}(\sigma\nu\rho\tau)$ means that when forming the elementary weight, the root introduces the symbol β_i, rather than b_i.

For the $\mathbf{q}$ components of the true solution, the expansion is

$$\mathbf{q} + h\mathbf{v} + \sum_{m=2}^{\infty} \frac{h^m}{m!} \sum_{\sigma\nu\rho\tau \in \mathcal{N}T_{m-1}} \alpha(\sigma\nu\rho\tau)\, f(\sigma\nu\rho\tau)(\mathbf{v},\mathbf{q}). \quad (4.15)$$

On comparing (4.14) with (4.15), we infer that to ensure $O(h^{r+1})$ local errors in the q variables the conditions

$$\tilde{\Phi}(\sigma\nu\rho\tau) = \frac{1}{(m+1)\gamma(\sigma\nu\rho\tau)} \quad (4.16)$$

should hold for all $\sigma\nu\rho\tau$ of order $m \leq r-1$. It is not difficult to show that (4.16) follows from (4.13) and (4.12) (see e.g. Hairer *et al.* (1987), Chapter II, Lemma 13.13).

Remark 4.3 If we do not assume that the weights β_i are linked to the b_i through (4.12), then to guarantee order r two sets of order conditions are required, namely:

1. Equation (4.13) for all $\sigma\nu\rho\tau$ of order $\leq r$.
2. Equation (4.16) for all $\sigma\nu\rho\tau$ of order $\leq r - 1$.

Then the situation is very close to that of PRK methods where there are also two sets of conditions, one corresponding to graphs with white root and the other corresponding to graphs with black root. □

Remark 4.4 Following up the preceding remark, let us consider the case of an RKN scheme induced by a PRK scheme via (3.33). We can study the order conditions of the algorithm either by seeing it as an RKN method applied to (3.30) or by seeing it as a PRK method applied to (3.29), a particular case of partitioned system (3.23). In the PRK approach one should take into account that since $\mathbf{g}$ is the identity map, $\mathbf{g}(\mathbf{p}) \equiv \mathbf{p}$, second and higher partial derivatives of $\mathbf{g}$ vanish, so that some elementary differentials vanish and do not contribute to the local error. As a result, it is necessary to consider only bicolour rooted trees where black vertices have at most one son (this corresponds to meagre vertices having at most one son in the RKN case). It is then a simple exercise to check that the sets of RKN conditions 1 and 2 in the preceding remark respectively coincide with the PRK order conditions arising from white-rooted and black-rooted bicolour trees.

For these reasons, the order of a PRK method can be strictly lower than the order of the RKN method it induces; for the former there are more order conditions to be satisfied as black vertices are allowed to have more than one son. □

CHAPTER 5

Implementation

5.1 Variable step sizes

The easiest way to implement a one-step method Ψ is with constant step sizes, i.e., by prescribing a value of h and then setting $t_n = t_0 + nh$. However such implementations tend to be inefficient and, in general, it is better to vary h_{n+1} as the integration proceeds, in order to use small step sizes when the solution $y(t)$ changes rapidly and large step sizes when the solution is only slowly varying. In fact, it is possible to adjust h_{n+1} so as to ensure that at each step the norm of the local error (3.4) is below a prescribed *tolerance TOL* (Hairer *et al.* (1987), Section II.4). To this end, the method Ψ in (3.2) is supplemented by a second method $\hat{\Psi}$ of order $\hat{r} < r$ (usually $\hat{r} = r - 1$ or $r - 2$). Since Ψ is of order r, it holds that, for the local error $\Phi - \hat{\Psi}$ in the $\hat{\Psi}$ method (Φ is the true solution operator),

$$\begin{aligned}
&\Phi_\mathbf{F}(t_{n+1}, t_n)\mathbf{y}^n - \hat{\Psi}_\mathbf{F}(t_{n+1}, t_n)\mathbf{y}^n \\
&= \Psi_\mathbf{F}(t_{n+1}, t_n)\mathbf{y}^n - \hat{\Psi}_\mathbf{F}(t_{n+1}, t_n)\mathbf{y}^n \\
&\quad + \Phi_\mathbf{F}(t_{n+1}, t_n)\mathbf{y}^n - \Psi_\mathbf{F}(t_{n+1}, t_n)\mathbf{y}^n \\
&= \Psi_\mathbf{F}(t_{n+1}, t_n)\mathbf{y}^n - \hat{\Psi}_\mathbf{F}(t_{n+1}, t_n)\mathbf{y}^n + O(h_{n+1}^{r+1}).
\end{aligned}$$

Therefore

$$\mathbf{EST}_{n+1} = \Psi_\mathbf{F}(t_{n+1}, t_n)\mathbf{y}^n - \hat{\Psi}_\mathbf{F}(t_{n+1}, t_n)\mathbf{y}^n \qquad (5.1)$$

differs from the $O(h_{n+1}^{\hat{r}+1})$ local error of the $\hat{\Psi}$ formula in higher-order terms $O(h_{n+1}^{r+1})$ and can be used to estimate that local error.

If the norm of the estimate (5.1) computed at time t_{n+1} is below TOL, then the step from t_n to t_{n+1} is accepted and one proceeds to compute the approximation to the solution at time t_{n+2}. If the norm of the estimate exceeds the tolerance, then the approximation at t_{n+1} is rejected and the step from t_n to t_{n+1} is attempted again with a smaller step size. In either case, the value of the next step size to be employed is given by the formula

$$h_c(TOL/\|\mathbf{EST}\|)^{1/(\hat{r}+1)}, \qquad (5.2)$$

where h_c denotes the current value of the step size. The expression (5.2) arises as follows. If the $O(h_{n+1}^{\hat{r}+1})$ norm of the local error were exactly proportional to $h_{n+1}^{\hat{r}+1}$, then (5.2) would be the largest step size that leads to a norm of the local error below the tolerance. In practice (5.2) is multiplied by a safety factor, say 0.8, to decrease the probability of a rejection at the next step. Note that (5.2) provides h_{n+2} if the step $t_n \to t_{n+1}$ was accepted, but corresponds again to h_{n+1} when the step is going to be reattempted after a rejection.

When the step $t_n \to t_{n+1}$ has been completed, two numerical approximations at time t_{n+1}, $\Psi_{\mathbf{F}}(t_{n+1}, t_n)\mathbf{y}^n$ and $\hat{\Psi}_{\mathbf{F}}(t_{n+1}, t_n)\mathbf{y}^n$, are available. It is then possible to choose one or the other as the value of $\mathbf{y}^{n+1}$ to be employed to begin the new step from t_{n+1} to t_{n+2}. Often, the more accurate, higher-order result given by Ψ is chosen; this is referred to as *local extrapolation*. In what follows, we invariably assume local extrapolation.

In practice it may be advisable to scale the i-th component of the estimator (5.1) by a factor d_i before taking norms and comparing with TOL. The choice where d_i is the absolute value of the i-th component of the solution vector $\mathbf{y}^{n+1}$ corresponds to relative errors. Of course, the choice $d_i = 1$ corresponds to absolute errors.

The reader is directed to the survey paper by Shampine and Gladwell (1992) and to the references therein for more details on the construction of variable step size algorithms.

5.2 Embedded pairs

If the basic method Ψ is an RK method with tableau (3.11), then the auxiliary $\hat{\Psi}$ used for step size control is invariably chosen to be an RK method. The efficiency of the pair is enhanced if the coefficients $\hat{a}_{ij}$ of $\hat{\Psi}$ coincide with the corresponding a_{ij}. In this way the stage vectors $\mathbf{Y}_i$ are, at each step, the same for both methods and to find $\hat{\mathbf{y}}^{n+1}$, once the main $\mathbf{y}^{n+1}$ is available, we just have to compute a linear combination of known vectors

$$\hat{\mathbf{y}}^{n+1} = \mathbf{y}^n + h_{n+1}\sum_{i=1}^{s}\hat{b}_i\mathbf{F}(\mathbf{Y}_i, t_n + c_i h_{n+1}).$$

With the local extrapolation formulation we are assuming, the

EMBEDDED PAIRS

low-order approximation $\hat{y}^{n+1}$ is not actually required; it is sufficient to form the estimator

$$h_{n+1} \sum_{i=1}^{s} (b_i - \hat{b}_i) \mathbf{F}(\mathbf{Y}_i, t_n + c_i h_{n+1}).$$

A pair of RK formulae that share the same stage vectors is called an *embedded pair*.
Similar considerations of course apply for PRK and RKN methods. In the Nyström case, Ψ and $\hat{\Psi}$ share the same γ_i and α_{ij}. Embedded pairs of RK or RKN formulae are standard tools in the numerical integration of differential systems.

Example 5.1 Dormand *et al.* (1987a), (1987b) have optimized the construction of embedded RKN pairs. The following tableau gives the coefficients of an order-4 method with an embedded $\hat{\Psi}$ of order 3:

0				
$\frac{1}{4}$	$\frac{1}{32}$			
$\frac{7}{10}$	$\frac{7}{1000}$	$\frac{119}{500}$		
1	$\frac{1}{14}$	$\frac{8}{27}$	$\frac{25}{189}$	
	$\frac{1}{14}$	$\frac{8}{27}$	$\frac{25}{189}$	0
	$\frac{1}{14}$	$\frac{32}{81}$	$\frac{250}{567}$	$\frac{5}{54}$
	$-\frac{7}{150}$	$\frac{67}{150}$	$\frac{3}{20}$	$-\frac{1}{20}$
	$\frac{13}{21}$	$-\frac{20}{27}$	$\frac{275}{189}$	$-\frac{1}{3}$

Here the column vector on the left provides the abscissae γ_i for both Ψ and $\hat{\Psi}$; the 4 × 4 matrix gives the α_{ij} coefficients for both methods (elements not explicitly displayed are 0); the four row vectors at the bottom respectively provide the β_i for Ψ, the b_i for Ψ, the $\hat{\beta}_i$ for $\hat{\Psi}$ and the $\hat{b}_i$ for $\hat{\Psi}$.

A survey of the ideas used in the construction of modern, optimized RK and RKN formulae is given by Dormand and Prince (1989). These optimized methods have superseded the formulae derived by Runge, Kutta and other classical authors and often found in textbooks. An example of the application of the optimization techniques is given in Subsection 8.5.3. □

Remark 5.1 Note that in the tableau of the example above, $\gamma_4 = 1$ and, for each i, $\alpha_{4i} = \beta_i$. These relations imply that the last stage vector Q_4 in the current step coincides with the approximation q^{n+1} at the end of the step. Hence, the last f evaluation in the current step coincides with the first evaluation to be used in the next step, a property known as *FSAL (first same as last)*. This property entails that, even though we are dealing with a four-stage method, only three function evaluations are required at each step (other than the very first $t_0 \to t_1$). Many standard methods are based on the FSAL idea. □

5.3 Numerical experience with variable step sizes

We have implemented, with variable step sizes, the RKN embedded pair in Example 5.1 by following the general pattern of the program Doprin in the book by Hairer et al. (1987). We have also implemented with constant step sizes order-4, FSAL method of the same embedded pair.

As a test example we have chosen Kepler's problem in cartesian coordinates (p_1, p_2, q_1, q_2) (see Subsection 1.2.4), with the initial conditions (1.15). The ratio of the distances r to the origin of the apocentre and pericentre is $(1 + e)/(1 - e)$. When the eccentricity e is close to 1, this ratio is large. Conservation of angular momentum $r^2 \dot\theta$ implies that $\dot\theta$ is large in the pericentre and small in the apocentre. As a consequence, it is appropriate to use small step sizes near the pericentre and large step sizes near the apocentre. In fact, Kepler's system with initial condition (1.15) is often used as a test problem for variable step size integrators (see e.g. Dormand et al. (1987a)).

Fig. 5.1 gives the error at time $30 \times 2\pi$ (i.e., after 30 periods) as a function of the CPU time in seconds on a SUN Sparc IPX workstation. The eccentricity is $e = 0.3$. The error is measured in the standard (Euclidean) norm of the space $\mathcal{R}^4$ of the variables (p_1, p_2, q_1, q_2). The solid line corresponds to the variable step size code. There are seven runs corresponding to values of the tolerance (absolute error) varying from 10^{-4} to 10^{-10}. The dotted line corresponds to the constant step size, order-4 method. There are six runs corresponding to $h = 2\pi/64, 2\pi/128, \ldots, 2\pi/2048$. We see that the variable step size implementation is more expensive; the CPU times needed for both algorithms to achieve a given error are in a ratio of approximately 3 to 2.

Fig. 5.2 is similar to Fig. 5.1, but now $e = 0.5$. (There are also fewer runs, now the tolerance starts at 10^{-5} and h starts at $2\pi/128$. Runs leading to errors larger than 0.1 have not been included.) The

NUMERICAL EXPERIENCE WITH VARIABLE STEP SIZES

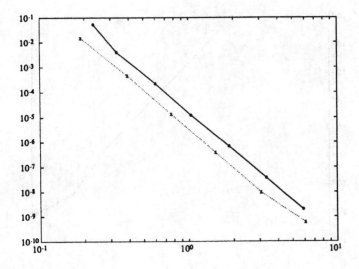

Figure 5.1. *Error as a function of CPU time, $e = 0.3$*

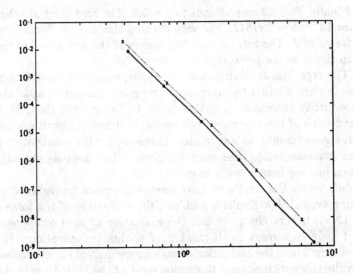

Figure 5.2. *Error as a function of CPU time, $e = 0.5$*

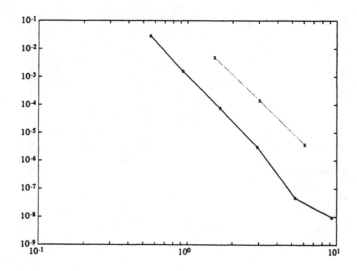

Figure 5.3. *Error as a function of CPU time, e = 0.7*

variable step size implementation is now slightly cheaper than the constant step size algorithm.

Finally, Fig. 5.3 corresponds to $e = 0.7$. The runs start at tolerance 10^{-5}, $h = 2\pi/512$. The variable step size code is cheaper by a factor of 2. The ratio of the step size near the apocentre to the step size near the pericentre is approximately 22.

The experiments clearly show that, as expected, the variable step size implementation becomes more and more advantageous as the eccentricity increases. A further point to be noted is that, for a fixed value of the tolerance in the variable step size algorithm, the error grows mildly as e increases. However, for the constant step size implementation, the error for given h increases significantly when moving from $e = 0.5$ to $e = 0.7$.

In a sense, the results we have presented cannot be regarded as being typical. For Kepler's problem, the evaluation of the force **f** in (3.30) is very cheap; in fact the evaluation of each component of **f** only requires a small number of arithmetic operations. For problems where the evaluation of each component of **f** is expensive, (explicit) algorithms tend to employ most of the CPU time in the function evaluations; the time devoted to overheads such as forming the stage vectors can be neglected. It is then a standard practice

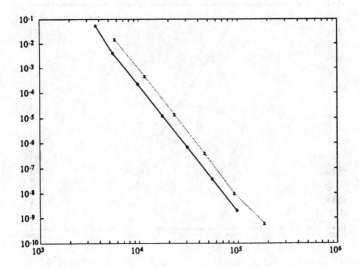

Figure 5.4. *Error as a function of number of evaluations, $e = 0.3$*

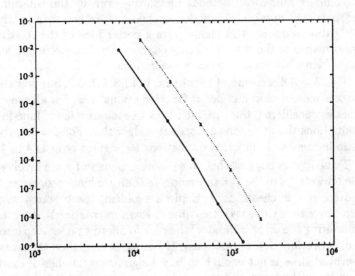

Figure 5.5. *Error as a function of number of evaluations, $e = 0.5$*

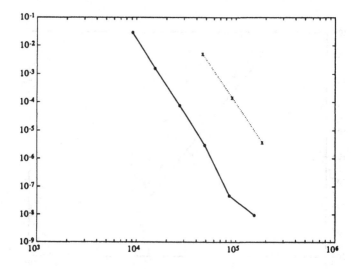

Figure 5.6. *Error as a function of number of evaluations*, $e = 0.7$

to compare numerical methods measuring work by the (machine independent) number of function evaluations, rather than by the CPU time required; this should give a better idea of the relative performance of the methods being compared if these were applied to problems involving expensive evaluations.

Figs. 5.4–5.6 correspond to the runs in Figs. 5.1–5.3, but now the work is measured in number of function evaluations. For all eccentricities considered, the variable step size code uses fewer function evaluations than the constant step size algorithm. For $e = 0.7$, the ratio in number of function evaluations for a given error is 4 to 1.

These figures suggest that when more expensive f's are involved the advantages of the variable step size code are more pronounced. Indeed, if f is cheap, as in Kepler's problem, the constant step size implementation uses the time it saves in overheads (such as computing the error estimator) in order to afford smaller step sizes (i.e., more function evaluations). If f is expensive, then the saved overhead time is not enough to buy a significant number of extra function evaluations, and the constant step size algorithms become quite inefficient.

5.4 Implementing implicit methods

5.4.1 Reformulation of the equations

We now turn to implementation details of implicit methods. We begin with the RK case. Our treatment follows closely the material in Hairer and Wanner (1991), Chapter IV.8.

When solving (3.14) in practice it may be advisable, in order to reduce the influence of round-off errors, to consider as unknowns the increments

$$Z_i = Y_i - y^n.$$

Then the algebraic equations (3.14) to be solved for the stage vectors become

$$Z_i = h_{n+1} \sum_{j=1}^{s} a_{ij} F(y^n + Z_j, t_n + c_j h_{n+1}), \quad i = 1, \ldots, s. \quad (5.3)$$

When a solution Z_i, $i = 1, \ldots, s$ of this system has been found, then a direct application of (3.12) to find y^{n+1} requires s additional evaluations of F. These can be avoided if the matrix $A = (a_{ij})$ of the RK coefficients is invertible. In fact (5.3) can be rewritten as

$$\begin{bmatrix} Z_1 \\ \vdots \\ Z_s \end{bmatrix} = h_{n+1}(A \otimes I) \begin{bmatrix} F(y^n + Z_1, t_n + c_1 h_{n+1}) \\ \vdots \\ F(y^n + Z_s, t_n + c_s h_{n+1}) \end{bmatrix},$$

so that (3.12) becomes

$$y^{n+1} = y^n + \sum_{i=1}^{s} d_i Z_i,$$

with

$$(d_1, \ldots, d_s) = (b_1, \ldots, b_s) A^{-1}.$$

Above, $\otimes$ denotes the tensor (Kronecker) product of matrices (see e.g. Dekker and Verwer (1984), Section 3.7),

$$A \otimes I = \begin{bmatrix} a_{11} I & \cdots & a_{1s} I \\ \cdots & \cdots & \cdots \\ a_{s1} I & \cdots & a_{ss} JI \end{bmatrix}.$$

Example 5.2 For the Gauss method of order 4 we find $d_1 = -\sqrt{3}$, $d_2 = +\sqrt{3}$. □

5.4.2 Solving the equations: functional iteration

Functional iteration provides the most straightforward technique to solve (5.3). With superscripts denoting iteration number, we have, for $\nu = 0, 1, \ldots$,

$$\mathbf{Z}_i^{[\nu+1]} = h_{n+1} \sum_{j=1}^{s} a_{ij} \mathbf{F}(\mathbf{y}^n + \mathbf{Z}_j^{[\nu]}, t_n + c_j h_{n+1}), \quad i = 1, \ldots, s.$$

5.4.3 Solving the equations: Newton-like iteration

It is well known that the procedure described in the preceding subsection works if the right-hand side of (5.3) is a contractive mapping. This requires that h_{n+1} be small with respect to the inverse of the Lipschitz constant L of $\mathbf{F}$. As commented before, implicit methods have only been applied in practice to stiff problems. Since in these L is large, the use of functional iteration is out of the question and it has been standard to resort to (simplified) Newton iterations.

The application of Newton's method to (5.3) involves the solution of linear systems with matrices of the form

$$\begin{bmatrix} I - h_{n+1} a_{11} J_1^{[\nu]} & \cdots & -h_{n+1} a_{1s} J_s^{[\nu]} \\ \cdots & \cdots & \cdots \\ -h_{n+1} a_{s1} J_1^{[\nu]} & \cdots & I - h_{n+1} a_{ss} J_s^{[\nu]} \end{bmatrix},$$

where $J_i^{[\nu]}$ denotes the Jacobian matrix of $\mathbf{F}$ with respect to $\mathbf{y}$ evaluated at $(\mathbf{y}^n + \mathbf{Z}_i^{[\nu]}, t_n + c_i h_{n+1})$.

To simplify things, all Jacobians $J_i^{[\nu]}$ are replaced by an approximation

$$J \approx \frac{\partial \mathbf{F}}{\partial \mathbf{Y}}(\mathbf{y}^n, t_n).$$

Note that, in particular, this implies that the Jacobians do not change with ν, i.e., they are not updated within each time step. The simplified Newton iteration for (5.3) reads

$$(I - h_{n+1} A \otimes J) \Delta \mathcal{Z}^{[\nu]} = -\mathcal{Z}^{[\nu]}$$
$$+ h_{n+1}(A \otimes I)\mathcal{F}(\mathcal{Z}^{[\nu]}), \quad (5.4)$$

$$\mathcal{Z}^{[\nu+1]} = \mathcal{Z}^{[\nu]} + \Delta \mathcal{Z}^{[\nu]}. \quad (5.5)$$

Here $\mathcal{Z}^{[\nu]}$ is the block vector consisting of s D-dimensional blocks $\mathbf{Z}_i$ and $\mathcal{F}(\mathcal{Z}^{[\nu]})$ is the block vector of s D-dimensional blocks $\mathbf{F}(\mathbf{y}^n + \mathbf{Z}_i^{[\nu]}, t_n + c_i h_{n+1})$. Each iteration requires the solution of an $s \times D$ linear system. The matrix of the system $I - h_{n+1}A \otimes J$ does not change from one iteration to the next and hence its factorization is done only once. This factorization, requiring $O((sD)^3)$ operations, represents the bulk of the cost of the linear algebra.

Remark 5.2 Savings in the linear algebra are possible (Hairer and Wanner (1991), Chapter IV.8). If A is invertible, then a technique due to Butcher and Bickart premultiplies (5.4) by $(h_{n+1}A)^{-1} \otimes I$ and transforms A^{-1} into a matrix Λ with a simpler structure (for instance diagonal, block diagonal, triangular)

$$T^{-1}AT = \Lambda.$$

In the transformed variables

$$\mathcal{W} = (T^{-1} \otimes I)\mathcal{Z}$$

the iteration (5.4)–(5.5) reads

$$\begin{aligned}(h_{n+1}^{-1}\Lambda \otimes I - I \otimes J)\Delta\mathcal{W}^{[\nu]} &= -h_{n+1}^{-1}(\Lambda \otimes I)\mathcal{W}^{[\nu]} \\ &\quad + (T^{-1} \otimes I)\mathcal{F}((T \otimes I)\mathcal{W}^{[\nu]}), \\ \mathcal{W}^{[\nu+1]} &= \mathcal{W}^{[\nu]} + \Delta\mathcal{W}^{[\nu]}.\end{aligned}$$

In the case where Λ is a diagonal matrix with diagonal entries λ_i the matrix $(h_{n+1}^{-1}\Lambda \otimes I - I \otimes J)$ to be factorized is now block diagonal with s blocks $h_{n+1}^{-1}\lambda_i I - J$. Hence now it is only necessary to factorize s $D \times D$ matrices with an operation count $sO(D^3)$. This is to be compared with the factorization of an sD-dimensional matrix in (5.4) with a count $O((sD)^3)$. In a like manner, savings occur if Λ, without being diagonal, has a simpler structure than A^{-1}.

Additional implementation ideas can be found in Cooper and Vignesvaran (1990) and references therein. □

5.4.4 Starting the iterations

With functional or Newton iteration, the initial $\mathbf{Z}_i^{[0]}$ can be determined in various ways. If the RK method is a collocation method, then a choice that suggests itself is as follows. Let $\mathbf{u}_{n-1}(t)$ be the collocation polynomial associated with the already completed step from t_{n-1} to t_n (see Subsection 3.3.2). Consider the polynomial of degree s

$$\mathbf{w}_{n-1}(t) = \mathbf{u}_{n-1}(t) - \mathbf{y}^n$$

that vanishes at t_n and, according to (3.15) (with n replaced by $n-1$), takes at the points $t_{n-1} + c_i h_n$ the values $\mathbf{Z}_i$ corresponding to the previous step. We use $\mathbf{w}_{n-1}(t)$ for t outside the interval $[t_{n-1}, t_n]$ to define the $\mathbf{Z}_i^{[0]}$ as the values of $\mathbf{w}_{n-1}(t)$ at the points $t_n + c_i h_{n+1}$.

5.4.5 Stopping the iterations

With functional or simplified Newton iteration, the convergence of $\mathcal{Z}^{[\nu]}$ is linear, i.e.

$$\|\Delta \mathcal{Z}^{[\nu+1]}\| \leq \Theta \|\Delta \mathcal{Z}^{[\nu]}\|,$$

with $\Theta < 1$. Standard results about fixed point iteration show that then, if $\mathcal{Z}$ denotes the exact solution of the algebraic system (5.3) being solved, then

$$\|\mathcal{Z}^{[\nu+1]} - \mathcal{Z}\| \leq \frac{\Theta}{1-\Theta} \|\Delta \mathcal{Z}^{[\nu]}\|.$$

The convergence factor Θ can be estimated by the computed quantities

$$\Theta_\nu = \frac{\|\Delta \mathcal{Z}^{[\nu]}\|}{\|\Delta \mathcal{Z}^{[\nu+1]}\|}, \quad \nu = 1, 2, \ldots$$

Therefore the iterations can be stopped when

$$\eta_\nu \|\Delta \mathcal{Z}^{[\nu]}\| \leq TOL, \quad \eta_\nu = \frac{\Theta_\nu}{1 - \Theta_\nu},$$

where TOL is a given tolerance. This only works after two iterations. To be able to stop after just one iteration, we define for $\nu = 0$ the quantity $\eta_0 = \eta_{old}^{0.8}$, where η_{old} is the last ν in the preceding step.

Remark 5.3 The values Θ_ν can be used to implement savings in the case of Newton iteration. If Θ_ν is small, say below 10^{-3}, we do not recompute and factorize the Jacobian at the next time step. □

5.4.6 The algebraic equations in the Runge-Kutta-Nyström case

Considerations similar to those above apply to implicit RKN formulae. When rewriting the equations (3.32), for the stages we consider as new unknowns

$$\mathbf{Z}_i = \mathbf{Q}_i - \mathbf{q}^n - h_{n+1} \gamma_i \mathbf{v}^n.$$

In terms of these, (3.32) becomes

$$Z_i = h_{n+1}^2 \sum_{j=1}^{s} \alpha_{ij} f(q^n + h_{n+1}\gamma_j v^n + Z_j, t_n + \gamma_j h_{n+1}),$$

a system that can be solved by functional or simplified Newton iterations. Note now the h_{n+1}^2 factor in the right-hand side; this is to be compared with the h_{n+1} factor in (5.3).

5.5 Numerical experiments with the fourth-order Gauss method

We have implemented, with constant step sizes, the two-stage Gauss method. Four algorithms are considered. Two of them use the method as a standard RK procedure (Subsection 3.3.2) applied to a first-order system. The other two use the method in an RKN formulation, see Example 3.3. For each formulation, both functional iteration and (simplified) Newton iteration are considered. The ideas above have been used, with the only exception that the linear algebra savings in Remark 5.2 are not implemented: the matrix A of the coefficients is 2×2 with complex eigenvalues and hence not amenable to much simplification by similarity transformations.

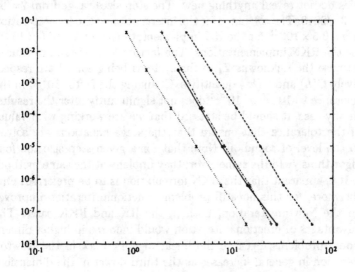

Figure 5.7. *Error as a function of CPU time*, $e = 0.5$

Table 5.1. *Functional iteration runs*

h	RK iterations per step	RKN iterations per step
$2\pi/64$	9.6	5.4
$2\pi/128$	7.5	4.1
$2\pi/256$	5.9	3.5
$2\pi/512$	4.9	3.2
$2\pi/1024$	4.3	2.7
$2\pi/2048$	3.5	2.4

The four algorithms have been applied to Kepler's problem as in Section 5.3. Note that in the Newton case, we solve 8 × 8 systems in the RK formulation and 4 × 4 systems in the RKN formulation.

In Fig. 5.7, the leftmost dotted line corresponds to the RKN formulation with functional iteration, the solid line in the middle to the RKN formulation with Newton iteration, the dash–dot line to the RK version with functional iteration and, finally, the rightmost broken line to the RK format with Newton iteration.

The figure corresponds to $e = 0.5$; the results for other eccentricities do not reveal anything new. The step sizes range from $2\pi/64$ to $2\pi/2048$. The tolerance for the interruption of the inner iteration is $h \times 0.5 \times 10^{-15}$ for the RK implementations and $h^2 \times 0.5 \times 10^{-15}$ for the RKN implementations. The factors h or h^2 are introduced because the unknowns $\mathbf{Z}_i$ in the systems being solved are respectively $O(h)$ and $O(h^2)$ quantities. Changing the factor 10^{-16} in the tolerance to 10^{-15} or 10^{-14} does not significantly alter the results. In any case, it should be stressed that we are working with values of the tolerance that ensure that the stage equations are solved to the level of round-off. Note that for a given step size the four algorithms yield the same error; they implement the same method.

It is apparent that the RKN formulation is to be preferred. Furthermore, for this nonstiff problem, functional iteration improves on the Newton iteration, both in the RK and RKN cases. The advantages of functional iteration would increase in higher dimensional problems, because the linear algebra work in the Newton iteration in general increases as the third power of the dimension. For the same reason, the advantages of functional iterations would be more marked for Gauss methods with more than two stages.

FOURTH-ORDER GAUSS METHOD

Table 5.2. *Statistics for the Newton iteration runs*

h	RK Iterations per step	RK Jacobian evaluations	RKN Iterations per step	RKN Jacobian evaluations
$2\pi/64$	4.8	350	–	–
$2\pi/128$	4.4	245	5.0	125
$2\pi/256$	4.1	119	4.1	3
$2\pi/512$	3.6	5	3.4	1
$2\pi/1024$	3.1	1	3.0	1
$2\pi/2048$	3.0	1	3.0	1

Table 5.1 provides additional insight into the computer runs in Fig. 5.7 with functional iteration. As expected, the number of iterations per step decreases with decreasing h: a smaller h means more accurate initial guesses $Z_i^{[0]}$ and also a smaller Lipschitz constant in the system. The RKN format, where the system has an $O(h^2)$ Lipschitz constant, requires fewer iterations per step than the RK format with an $O(h)$ Lipschitz constant.

Table 5.2 corresponds to the simplified Newton implementations. For small values of h the algorithms do not refresh the Jacobian throughout the integration (cf. Remark 5.3). A comparison with Table 5.1 shows that, in the RKN formulation, even for larger values of h (with the Jacobian updated from time to time) the number of Newton iterations is not smaller than the number of iterations required by functional iteration. This is of course the effect of inaccurate Jacobians due to the technique described in Remark 5.3.

Remark 5.4 We also performed experiments with the simplified Newton iteration where the Jacobian was recomputed at every step, thus overruling Remark 5.3. As expected, this reduces substantially the number of iterations per step, both for the RK and RKN versions.

In the RKN version, the reduction in iterations makes up for the cost of computing and factorizing more Jacobians and therefore the CPU times become smaller than those reported in Fig. 5.7. However functional iteration is still more effective for all values of h tried.

In the RK version, with larger systems, the reduction in number of iterations due to accurate Jacobians does not compensate for the extra cost of computing and factorizing more Jacobians. □

CHAPTER 6

Symplectic integration

6.1 Symplectic methods

In Chapter 2 we saw that the distinctive feature of the solution operator Φ of Hamiltonian problems is symplecticness (or in the one-degree-of-freedom case conservation of oriented area). In Chapter 3 we examined how numerical methods replace Φ by an approximation Ψ. If we wish the approximation Ψ to retain the 'Hamiltonian' features of Φ, we should insist on Ψ also being a symplectic transformation. However, most standard numerical integrators replace Φ by a *nonsymplectic* mapping Ψ. This is illustrated in Fig. 6.1

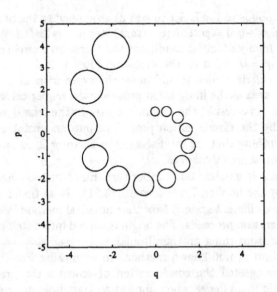

Figure 6.1. *The harmonic oscillator integrated by the explicit Euler method*

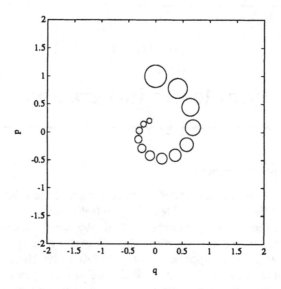

Figure 6.2. *The harmonic oscillator integrated by the implicit Euler method*

that corresponds to the Euler rule (3.3) as applied to the harmonic oscillator $\dot{p} = -q$, $\dot{q} = p$. The (constant) step size is $2\pi/12$. We have taken as a family of initial conditions the points on a circle centred at $p = 1$, $q = 0$ and seen the evolution after 1, 2, ..., 12 steps. Clearly the circle, which should move clockwise without changing area, gains area as the integration proceeds: the numerical Ψ is not symplectic. As a result, the origin, a centre in the true dynamics, is turned by the discretization procedure into an unstable spiral point, something that, as we discussed in Chapter 2, cannot arise in real Hamiltonian dynamics.

Figure 6.2 is similar, but the forward Euler method has been replaced by the implicit Euler formula (3.21). (Note that Figs. 6.1 and 6.2 have different scales.) Now the numerical method loses area as the integration proceeds. The origin is turned into a stable spiral point, again something un-Hamiltonian.

This failure of well-known methods in mimicking Hamiltonian dynamics suggested the consideration of schemes that generate a symplectic mapping Ψ when applied to Hamiltonian problems. Such methods are called *symplectic* or *canonical*. Early references on symplectic integration are Ruth (1983), Channel (1983), Men-

SYMPLECTIC METHODS

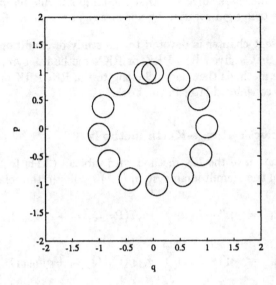

Figure 6.3. *The harmonic oscillator integrated by the implicit midpoint rule*

yuk (1984), Feng (1985), (1986a), (1986b), even though the idea of symplectic integration apparently goes back to DeVogelaere in 1956 (cf. Channell and Scovel (1990)). This book is primarily devoted to the study of symplectic integrators.

A simple example of a symplectic integrator is given by the midpoint rule (3.18). Figure 6.3 is similar to Figs. 6.1 and 6.2 but now the integration method is the midpoint rule. Clearly the initial circle is moved by the method without changing its area. The origin correctly remains a centre after discretization. The picture is qualitatively correct, even though there is a quantitative error: after 12 steps of length $2\pi/12$ the circle should have returned to its original location.

Remark 6.1 The fact that in Fig. 6.3 the circle moves without changing shape is not inherent to symplecticness: it is rather due to the test problem and method being used. We noticed in Example 2.2 that, for the harmonic oscillator (1.5) *with $m\omega = 1$*, the true flow is a rigid rotation and hence preserves the shape of figures. (Shape is of course not preserved in general; even the harmonic oscillator flow does not preserve it as soon as $m\omega \neq 1$ (cf. Fig.

2.1).) It furthermore turns out that the midpoint rule happens to replace the true rigid rotation by another rigid rotation. □

The present chapter is devoted to the study of conditions that guarantee that a given RK, PRK or RKN method is symplectic. Symplectic methods that do not belong to the RK, PRK or RKN classes are considered later in the book.

6.2 Symplectic Runge-Kutta methods

The application of the RK method with tableau (3.11) to the integration of the Hamiltonian system (1.1) results in the relations

$$\mathbf{P}_i = \mathbf{p}^n + h_{n+1} \sum_{j=1}^{s} a_{ij} \mathbf{f}(\mathbf{P}_j, \mathbf{Q}_j, t_n + c_j h_{n+1}), \quad (6.1)$$

$$\mathbf{Q}_i = \mathbf{q}^n + h_{n+1} \sum_{j=1}^{s} a_{ij} \mathbf{g}(\mathbf{P}_j, \mathbf{Q}_j, t_n + c_j h_{n+1}), \quad (6.2)$$

$$\mathbf{p}^{n+1} = \mathbf{p}^n + h_{n+1} \sum_{i=1}^{s} b_i \mathbf{f}(\mathbf{P}_i, \mathbf{Q}_i, t_n + c_i h_{n+1}), \quad (6.3)$$

$$\mathbf{q}^{n+1} = \mathbf{q}^n + h_{n+1} \sum_{i=1}^{s} b_i \mathbf{g}(\mathbf{P}_i, \mathbf{Q}_i, t_n + c_i h_{n+1}), \quad (6.4)$$

where $\mathbf{f}$ and $\mathbf{g}$ respectively denote the d-vectors with components $-\partial H/\partial q_i$ and $\partial H/\partial p_i$ and $\mathbf{P}_i$ and $\mathbf{Q}_i$ are the internal stages corresponding to the $\mathbf{p}$ and $\mathbf{q}$ variables. The following result was discovered independently by Lasagni (1988), Sanz-Serna (1988) and Suris (1988).

Theorem 6.1 *Assume that the coefficients of the method (3.11) satisfy the relations*

$$b_i a_{ij} + b_j a_{ji} - b_i b_j = 0, \quad i,j = 1, \ldots, s. \quad (6.5)$$

Then the method is symplectic.

Proof. We follow the technique used by Sanz-Serna (1988). Suris (1988) resorts to Jacobians rather than to differential forms. No proof is presented in Lasagni (1988).

We employ the notation

$$\mathbf{k}_i = \mathbf{f}(\mathbf{P}_i, \mathbf{Q}_i, t_n + c_i h_{n+1}), \quad \mathbf{l}_i = \mathbf{g}(\mathbf{P}_i, \mathbf{Q}_i, t_n + c_i h_{n+1})$$

for the 'slopes' at the stages. Differentiate (6.3)–(6.4) and form the exterior product to arrive at

$$d\mathbf{p}^{n+1} \wedge d\mathbf{q}^{n+1} = d\mathbf{p}^n \wedge d\mathbf{q}^n + h_{n+1} \sum_{i=1}^{s} b_i\, d\mathbf{k}_i \wedge d\mathbf{q}^n$$

$$+ h_{n+1} \sum_{j=1}^{s} b_j\, d\mathbf{p}^n \wedge d\mathbf{l}_j$$

$$+ h_{n+1}^2 \sum_{i,j=1}^{s} b_i b_j\, d\mathbf{k}_i \wedge d\mathbf{l}_j.$$

Our next step is to eliminate $d\mathbf{k}_i \wedge d\mathbf{q}^n$ and $d\mathbf{p}^n \wedge d\mathbf{l}_j$ from this expression. This is easily achieved by differentiating (6.1) and (6.2) and taking the exterior product of the result with $d\mathbf{k}_i$, $d\mathbf{l}_j$ respectively. The outcome of the elimination is

$$d\mathbf{p}^{n+1} \wedge d\mathbf{q}^{n+1} - d\mathbf{p}^n \wedge d\mathbf{q}^n$$

$$= h_{n+1} \sum_{i=1}^{s} b_i\, [d\mathbf{k}_i \wedge d\mathbf{Q}_i + d\mathbf{P}_i \wedge d\mathbf{l}_i]$$

$$- h_{n+1}^2 \sum_{i,j=1}^{s} (b_i a_{ij} + b_j a_{ji} - b_i b_j)\, d\mathbf{k}_i \wedge d\mathbf{l}_j. \qquad (6.6)$$

The second term in the right-hand side vanishes in view of (6.5). To finish the proof it is then sufficient to show that, for each i,

$$d\mathbf{k}_i \wedge d\mathbf{Q}_i + d\mathbf{P}_i \wedge d\mathbf{l}_i = 0. \qquad (6.7)$$

Dropping the subscript i that numbers the stages, we can write

$$d\mathbf{k} \wedge d\mathbf{Q} + d\mathbf{P} \wedge d\mathbf{l} = \sum_{\mu=1}^{d} [dk_\mu \wedge dQ_\mu + dP_\mu \wedge dl_\mu]$$

$$= \sum_{\mu,\nu=1}^{d} \left[\frac{\partial f_\mu}{\partial p_\nu} dP_\nu \wedge dQ_\mu + \frac{\partial f_\mu}{\partial q_\nu} dQ_\nu \wedge dQ_\mu \right.$$

$$\left. + \frac{\partial g_\mu}{\partial p_\nu} dP_\mu \wedge dP_\nu + \frac{\partial g_\mu}{\partial q_\nu} dP_\mu \wedge dQ_\nu \right].$$

To see that this expression vanishes, write f_μ and g_μ as derivatives of H

$$\frac{\partial f_\mu}{\partial p_\nu} = -\frac{\partial^2 H}{\partial p_\nu \partial q_\mu},$$

$$\frac{\partial f_\mu}{\partial q_\nu} = -\frac{\partial^2 H}{\partial q_\nu \partial q_\mu},$$

$$\frac{\partial g_\mu}{\partial p_\nu} = \frac{\partial^2 H}{\partial p_\nu \partial p_\mu},$$

$$\frac{\partial g_\mu}{\partial q_\nu} = \frac{\partial^2 H}{\partial p_\mu \partial q_\nu}.$$

This and the skew-symmetry of the wedge product imply

$$\sum_{\mu,\nu=1}^{d} \left[\frac{\partial f_\mu}{\partial p_\nu} dP_\nu \wedge dQ_\mu + \frac{\partial g_\mu}{\partial q_\nu} dP_\mu \wedge dQ_\nu \right] = 0,$$

$$\sum_{\mu,\nu=1}^{d} \frac{\partial f_\mu}{\partial q_\nu} dQ_\nu \wedge dQ_\mu = 0,$$

$$\sum_{\mu,\nu=1}^{d} \frac{\partial g_\mu}{\partial p_\nu} dP_\mu \wedge dP_\nu = 0.$$

This concludes the proof. □

Remark 6.2 The symplecticness of the method should be understood in the following sense. Assume that, for given t_n and h_{n+1},

$$(\mathbf{p}^{n+1}, \mathbf{q}^{n+1}) = \Psi_H(t_{n+1}, t_n)(\mathbf{p}^n, \mathbf{q}^n)$$

is a smooth function defined in a subdomain of Ω and satisfying the RK equations (6.1)–(6.4). Then $\Psi_H(t_{n+1}, t_n)$ is a symplectic transformation. This shows that both the numerical solution branch that approximates the theoretical solution and the spurious branches (see Section 3.3.3) will be symplectic. □

As we shall see in Section 6.5 below, the conditions (6.5) are also essentially necessary for the method to be symplectic. *In the rest of the book we therefore use the term 'symplectic RK methods' to refer to RK methods that satisfy (6.5).* Such methods are all *implicit:* after setting $j = i$ in (6.5) it is obvious that consistent, symplectic RK methods cannot be explicit.

Remark 6.3 Due to symmetry considerations, (6.5) holds if

$$b_i a_{ij} + b_j a_{ji} - b_i b_j = 0$$

for $i \leq j$ (or alternatively for $i \geq j$). Thus (6.5) really imposes $s(s+1)/2$ independent equations on the $s^2 + s$ elements of the RK tableau. □

Remark 6.4 It should be pointed out that the left-hand side of (6.5) provides the entries of the matrix that features in the definition of *algebraic stability* introduced, in connection with stiff systems, by Burrage and Butcher (1979) and Crouzeix (1979) (see Dekker and Verwer (1984)). □

Remark 6.5 There is no relation between the properties of an RK method being symmetric and being symplectic. It is easy to construct examples of two-stage methods that are symmetric but are not symplectic or that are symplectic but not symmetric. The midpoint rule provides an example of a method that is both symplectic and symmetric. The Euler rule is neither symmetric nor symplectic. □

Before we close this section, it is advisable to comment briefly on the special case of *linear problems*. The integration of an autonomous linear problem $\dot{y} = Ay$, A a constant matrix, by an RK method, results in a recursion

$$y^{n+1} = R(h_{n+1}A)y^n,$$

where R is the *stability function*, a rational function associated with the method (Hairer and Wanner (1991), Chapter IV.3). Different RK tableaux may give rise to the same R; the methods are then different but coincide when applied to autonomous linear problems (see an example in Section 14.3). In the Hamiltonian case, A is, according to (1.2), of the form $J^{-1}S$ with S a $2d \times 2d$ constant, symmetric matrix. The symplecticness condition (2.7) then becomes

$$R(h_{n+1}J^{-1}S)^T J\, R(h_{n+1}J^{-1}S) = J,$$

or

$$R(-h_{n+1}SJ^{-1})\, J\, R(h_{n+1}J^{-1}S) = J.$$

Now since for all positive integers ν

$$J(J^{-1}S)^\nu = (SJ^{-1})^\nu J,$$

we can further write

$$R(-h_{n+1}SJ^{-1})R(h_{n+1}SJ^{-1})J = J,$$

and we conclude that symplecticness *for linear autonomous* problems is equivalent to the requirement that for all h_{n+1} and symmetric S

$$R(-h_{n+1}SJ^{-1})R(h_{n+1}SJ^{-1}) = I.$$

Clearly this holds if and only if

$$R(-z)R(z) \equiv 1,$$

i.e., if and only if the stability function satisfies the symmetry requirement

$$R(z)^{-1} \equiv R(-z).$$

Note that the last identity implies that for purely imaginary z, $|R(z)| = 1$.

6.3 Symplectic Partitioned Runge-Kutta methods

Often the Hamiltonian function has the special structure

$$H(\mathbf{p}, \mathbf{q}, t) = T(\mathbf{p}) + V(\mathbf{q}, t). \tag{6.8}$$

In mechanics T and V would respectively represent the *kinetic* and *potential* energy. Hamiltonians of this form are called *separable*. All the examples in Chapter 1 have separable Hamiltonians (provided that the Kepler and modified Kepler problems are treated in cartesian coordinates; in polar coordinates the kinetic energy of a point mass is given by $(1/(2m))(p_r^2 + r^{-2} p_\theta^2)$ and depends on the radial coordinate r).

When the Hamiltonian is separable, the Hamilton equations take the partitioned form (3.23), with

$$\mathbf{f} = -\nabla_\mathbf{q} V, \qquad \mathbf{g} = \nabla_\mathbf{p} T,$$

and can be integrated by a PRK method like (3.24).

The following result was first given by Sanz-Serna at the London 1989 ODE meeting (see Sanz-Serna (1992a)) and discovered independently by Suris (1990). See also Abia and Sanz-Serna (1993).

Theorem 6.2 *Assume that the coefficients of the method (3.24) satisfy the relations*

$$b_i A_{ij} + B_j a_{ji} - b_i B_j = 0, \quad i, j = 1, \ldots, s. \tag{6.9}$$

Then the method is symplectic when applied to Hamiltonian problems with separable Hamiltonian (6.8).

Proof. It is similar to that of Theorem 6.1. Instead of (6.6) we now find

$$d\mathbf{p}^{n+1} \wedge d\mathbf{q}^{n+1} - d\mathbf{p}^n \wedge d\mathbf{q}^n$$

$$= h_{n+1} \sum_{i=1}^{s} [b_i \, d\mathbf{k}_i \wedge d\mathbf{Q}_i + B_i \, d\mathbf{P}_i \wedge d\mathbf{l}_i]$$

$$- h_{n+1}^2 \sum_{i,j=1}^{s} (b_i A_{ij} + B_j a_{ji} - b_i B_j) \, d\mathbf{k}_i \wedge d\mathbf{l}_j. \tag{6.10}$$

The second term in the right-hand side vanishes in view of (6.9). Furthermore, for each i, $d\mathbf{k}_i \wedge d\mathbf{Q}_i$ vanishes:

$$d\mathbf{k}_i \wedge d\mathbf{Q}_i = d\mathbf{f}(\mathbf{Q}_i, t_n + C_i h_{n+1}) \wedge d\mathbf{Q}_i, \qquad (6.11)$$

the Jacobian matrix of $\mathbf{f} = -\nabla_\mathbf{q} V$ is symmetric and the wedge product is skew-symmetric. Similarly $d\mathbf{P}_i \wedge d\mathbf{l}_i = 0$ and symplecticness follows. □

A comment similar to Remark 6.2 applies here. It is also true that (6.9) is also essentially necessary for the method to be symplectic (see Section 6.5). Therefore *we shall use the term 'symplectic PRK' as equivalent to the expression 'PRK method that satisfies (6.9)'*.

6.4 Symplectic Runge-Kutta-Nyström methods

A commonly occurring case of separable Hamiltonian function has $T = (1/2)\mathbf{p}^T M^{-1}\mathbf{p}$, with M a constant, symmetric, invertible matrix, so that the Hamiltonian reads

$$H(\mathbf{p}, \mathbf{q}, t) = \frac{1}{2}\mathbf{p}^T M^{-1}\mathbf{p} + V(\mathbf{q}, t). \qquad (6.12)$$

Often, M is diagonal and its diagonal entries represent the 'masses' in the system (mass matrix). All the examples in Chapter 1 have Hamiltonians of this form (provided that in the Kepler and modified Kepler problem cartesian coordinates are chosen). Hamiltonians that are separable but not of the form (6.12) arise in practice: a good example is the relativistic motion of a system of point masses (cf. Ruth (1983)). The relativistic kinetic energy of a point mass with momentum $\mathbf{p}$ and (rest) mass m is given by

$$T = T(\mathbf{p}) = c\sqrt{\mathbf{p}^T\mathbf{p} + m^2 c^2}$$

(c is the velocity of light).

For (6.12) the equations of motion are

$$\dot{\mathbf{p}} = \mathbf{f}(\mathbf{q}, t) = \nabla_\mathbf{q} V(\mathbf{q}, t), \qquad \dot{\mathbf{q}} = M^{-1}\mathbf{p} \qquad (6.13)$$

or, after elimination of $\mathbf{p}$,

$$\ddot{\mathbf{q}} = M^{-1}\mathbf{f}(\mathbf{q}, t),$$

a system that can be integrated by an RKN method. It is now convenient to write the equations that define the step in terms of the variables $\mathbf{p}$ and $\mathbf{q}$, rather than in terms of $\mathbf{v} = \dot{\mathbf{q}}$ and $\mathbf{q}$ as we did in our description of RKN methods in Section 3.5. For the stages the result is

$$Q_i = q^n + h_{n+1}\gamma_i M^{-1}p^n$$
$$+ h_{n+1}^2 \sum_{j=1}^{s} \alpha_{ij} M^{-1} f(Q_j, t_n + \gamma_j h_{n+1}), \qquad (6.14)$$

while the approximation at the next time level is given by

$$p^{n+1} = p^n + h_{n+1} \sum_{i=1}^{s} b_i f(Q_i, t_n + \gamma_i h_{n+1}),$$

along with

$$q^{n+1} = q^n + h_{n+1} M^{-1} p^n + h_{n+1}^2 \sum_{i=1}^{s} \beta_i M^{-1} f(Q_i + t_n + \gamma_i h_{n+1}).$$

The conditions for an RKN method to be canonical were first derived by Suris (1988), (1989).

Theorem 6.3 *Assume that the coefficients of the method (3.31) satisfy the conditions*

$$\beta_i = b_i(1 - \gamma_i), \quad i = 1, \ldots, s, \qquad (6.15)$$
$$b_i(\beta_j - \alpha_{ij}) = b_j(\beta_i - \alpha_{ji}), \quad i,j = 1, \ldots, s. \qquad (6.16)$$

Then the method is symplectic when applied to Hamiltonian problems with Hamiltonian of the form (6.12).

Proof. In the original proof, Suris used Jacobians. A proof based on differential forms, similar to those given above for the RK and PRK cases, was first carried out by Okunbor and Skeel (1992a), who only considered the situation where the mass matrix M is diagonal. Here we cater for nondiagonal M.

Before we address the proof proper, we derive a simple auxiliary result. Namely, we prove that if f and g are vector-valued functions and S a constant, symmetric matrix then

$$d(Sf) \wedge dg = df \wedge d(Sg). \qquad (6.17)$$

In fact we can respectively write, with subscripts denoting components,

$$d(Sf) \wedge dg = \sum_i \left(\sum_j s_{ij} df_j \right) \wedge g_i,$$

$$df \wedge d(Sg) = \sum_j f_j \wedge d\left(\sum_i s_{ji} g_i \right),$$

so that both sides of (6.17) have the common value

$$\sum_{ij} s_{ij} df_j \wedge dg_i.$$

Note that, in particular, (6.17) implies that

$$d(Sf) \wedge df = df \wedge d(Sf)$$

and this, with the skew-symmetry of the wedge product, leads to

$$d(Sf) \wedge df = df \wedge d(Sf) = 0. \qquad (6.18)$$

We are now ready for the proof of the theorem. Upon differentiating the equations that define $\mathbf{p}^{n+1}$ and $\mathbf{q}^{n+1}$ and taking the wedge product, we find (with $\mathbf{k}_i = \mathbf{f}(\mathbf{Q}_i, t_n + \gamma_i h_{n+1})$)

$$d\mathbf{p}^{n+1} \wedge d\mathbf{q}^{n+1}$$

$$= d\mathbf{p}^n \wedge d\mathbf{q}^n + h_{n+1} \sum_{i=1}^s b_i\, d\mathbf{k}_i \wedge d\mathbf{q}^n$$

$$+ h_{n+1} d\mathbf{p}^n \wedge d(M^{-1}\mathbf{p}^n) + h_{n+1}^2 \sum_{i=1}^s b_i\, d\mathbf{k}_i \wedge d(M^{-1}\mathbf{p}^n)$$

$$+ h_{n+1}^2 \sum_{i=1}^s \beta_i\, d\mathbf{p}^n \wedge d(M^{-1}\mathbf{k}_i)$$

$$+ h_{n+1}^3 \sum_{i,j=1}^s b_i \beta_j\, d\mathbf{k}_i \wedge d(M^{-1}\mathbf{k}_j).$$

The third term in the right-hand side vanishes in view of (6.18). In the second, we replace $\mathbf{q}^n$ by its value obtained from (6.14). The result, after taking (6.17) into account, is

$$d\mathbf{p}^{n+1} \wedge d\mathbf{q}^{n+1}$$

$$= d\mathbf{p}^n \wedge d\mathbf{q}^n + h_{n+1} \sum_{i=1}^s b_i\, d\mathbf{k}_i \wedge d\mathbf{Q}_i$$

$$+ h_{n+1}^2 \sum_{i=1}^s (b_i(1-\gamma_i) - \beta_i)\, d\mathbf{k}_i \wedge d(M^{-1}\mathbf{p}^n)$$

$$+ h_{n+1}^3 \sum_{i<j} [b_i(\beta_j - \alpha_{ij}) - b_j(\beta_i - \alpha_{ji})]\, d\mathbf{k}_i \wedge d(M^{-1}\mathbf{k}_j).$$

Now the second term in the right-hand side is zero by the argument following (6.11); the third and fourth terms vanish by (6.15)–(6.16). □

The condition (6.15) coincides with the standard *simplifying assumption* for RKN methods we found before in (4.12). A comment similar to Remark 6.2 applies here. Since the conditions (6.15)–(6.16) are essentially necessary for the method to be symplectic (see Section 6.5), *we shall use, in the remainder of the book, the expression 'symplectic RKN methods' to refer to RKN methods that satisfy (6.15)–(6.16)*. Also note that in (6.16) there are only $s(s-1)/2$ independent conditions; the conditions for $i = j$ are identities and the conditions for $i < j$ duplicate those for $i > j$.

6.5 Necessity of the symplecticness conditions

6.5.1 Preliminaries

In the preceding sections we have presented conditions that guarantee that a given RK, PRK or RKN method is symplectic. It is of obvious interest to ascertain whether those conditions are also necessary for symplecticness. Strictly speaking they are not, as the following example shows.

Example 6.1 Consider the RK method with tableau

$$\begin{array}{c|cc} \frac{1}{2} & 0 & 0 \\ \frac{1}{2} & 0 & 0 \\ \hline & \lambda & 1-\lambda \end{array}, \qquad (6.19)$$

where λ is any real constant. Since both rows of the matrix of coefficients a_{ij} are equal, it is clear that for this method the first and second stage vectors Y_1, Y_2, coincide for any differential system and any step length. Furthermore this common value of the stages is identical to the stage of the midpoint rule (3.18). Then the method (6.19) produces the same result y^{n+1} as the midpoint rule, which is symplectic according to (6.5). Thus (6.19) is symplectic. However it does not satisfy (6.5) if $\lambda \neq 1$. □

Admittedly (6.19) is a rather contrived example, obtained by writing a symplectic method in an artificially complicated way. Actually such contrived examples are the only ones that can be offered: we shall prove that the sufficient symplecticness conditions presented so far are also *necessary*, provided that the method has been *written without including redundant stages*. We consider first the PRK case and then the RK and RKN cases.

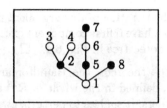

Figure 6.4. *Bicolour rooted tree for the proof of Theorem 6.4*

6.5.2 Independence of the elementary differentials in Hamiltonian problems

In the proof of the necessity of (6.9) we shall resort to the following result (Abia and Sanz-Serna (1993)), which is also of independent interest.

Theorem 6.4 *For each given bicolour rooted tree $\beta \rho \tau_0$ there is an autonomous, separable Hamiltonian system so that, if $[\mathcal{F}, \mathcal{G}](\beta \rho \tau)(0,0)$ denotes the elementary differential associated with $\beta \rho \tau$ evaluated at $(0,0)$, the following holds true. The first component of (the vector) $[\mathcal{F}, \mathcal{G}](\beta \rho \tau_0)((0,0)$ is $\neq 0$, while the first component of (the vector) $[\mathcal{F}, \mathcal{G}](\beta \rho \tau)(0,0)$ is 0 for all other bicolour rooted trees.*

Proof. Even though the proof is completely general, the underlying idea is best presented in an example. Let us consider the graph in Fig. 6.4, where the vertices have been labelled with the root having label 1. We set

$$H = p_2 p_3 p_4 + p_5 p_6 + p_7 + p_8 - q_1 q_2 q_5 q_8 - q_3 - q_4 - q_6 q_7.$$

There are as many terms in H as vertices in the graph. The term $p_2 p_3 p_4$ was introduced because we have a black vertex 2, with sons 3 and 4; the term $p_5 p_6$ originates from black vertex 5 with son 6, ..., the term $-q_1 q_2 q_5 q_8$ originates from white vertex 1 with sons 2, 5 and 8, etc. The differential system associated with this Hamiltonian function has (components of **f** and **g** are denoted by superscripts) $f^1 = q_2 q_5 q_8$ and $g^1 = 0$. At the origin, all derivatives of these two functions are zero, except for

$$\frac{\partial^3 f^1}{\partial q_2 \partial q_5 \partial q_8} = 1.$$

Hence, if an elementary differential at $(0, 0)$ has nonzero first component, the root of the corresponding graph must be white and have three sons. The iteration of this argument proves that all elementary differentials have 0 first component except that associated with the bicolour rooted tree in Fig. 6.4. □

Remark 6.6 In the theorem, the Hamiltonian function can be assumed to be C^∞, defined in the whole of $\mathcal{R}^{2d}$ and with bounded derivatives of all orders. It suffices to perturb, away from the origin, the polynomial Hamiltonian constructed in the proof. □

Remark 6.7 The theorem shows that the elementary differentials arising from (autonomous) separable Hamiltonian problems are *independent*. Hence, *within this class of problems*, for the Taylor expansions of the theoretical and numerical solution in Section 4.4 to agree up to $O(h^{r+1})$ terms, it is *necessary* that the order conditions (4.9) are fulfilled for all bicolour rooted trees of order $\leq r$. This reveals that the order of a PRK method when applied to the particular class of partitioned Hamiltonian problems is not higher than its conventional order (i.e., than its order applied to general, not necessarily Hamiltonian, problems). The same remark applies to RK schemes (because they can be thought of as a particular instance of PRK methods) and to RKN schemes (Calvo (1992)). □

6.5.3 Necessity of the symplecticness conditions in the Partitioned Runge-Kutta case

We are now in a position to prove, following Abia and Sanz-Serna (1993), the necessity of (6.9) when there are no *equivalent stages*. For RK methods, the notion of equivalent stages is carefully studied in Butcher (1987) Section 383. This material can easily be extended to the case of PRK methods; full details will not be given here. It is sufficient to mention that two stages $\mathbf{P}_i$ and $\mathbf{P}_j$ (or $\mathbf{Q}_k$ and $\mathbf{Q}_l$) of the method (3.24) are said to be equivalent if, to each smooth problem (3.23), each t_n and each initial point $(\mathbf{p}^n, \mathbf{q}^n)$ there corresponds a value h^* such that, when stepping from t_n to t_{n+1} with $h_{n+1} < h^*$, it holds that $\mathbf{P}_i = \mathbf{P}_j$ (or $\mathbf{Q}_k = \mathbf{Q}_l$). The equivalence of two stages can be characterized either in terms of the PRK tableaux or in terms of the *stage elementary weights*, see Butcher (1987) Section 383. The stage elementary weights are polynomials in the method coefficients; their definition is similar to that of the elementary weights of the method. For the stage $\mathbf{P}_k$ we replace by a_{ki} the b_i term corresponding to the root in the elementary weight of the method. For the stage $\mathbf{Q}_k$ we replace by A_{ki} the B_i term corresponding to the root in the elementary weight of the method.

Example 6.2 For $\beta\rho\tau_{2,1,w}$ in Fig. 4.3 we found in (4.10) that the elementary weight is given by $\sum_{i,j=1}^{s} b_i A_{ij}$. Hence the elementary weight of the stage $\mathbf{P}_k$ is $\sum_{i,j=1}^{s} a_{ki} A_{ij}$. Note that this is the $\beta\rho\tau_{2,1,w}$ elementary weight of the method obtained by replacing in the tableau of the given method the last row, b_i, $i = 1, \ldots, s$, by the row a_{ki}, $i = 1, \ldots, s$. □

The following result holds.

Lemma 6.5 *Assume that the PRK method (3.24) has no pair of equivalent stages. Then, there exists a C^∞ autonomous, separable Hamiltonian, defined in the whole of $\mathcal{R}^{2d}$ and with bounded partial derivatives of all orders, such that, at $t_0 = 0$ and $\mathbf{p}^0 = \mathbf{q}^0 = \mathbf{0}$, for all sufficiently small $h_1 > 0$ and all i, j, $i \neq j$, it holds that $\mathbf{P}_i \neq \mathbf{P}_j$ and $\mathbf{Q}_i \neq \mathbf{Q}_j$.*

Proof. By characterizing the equivalence of stages in terms of stage elementary weights, as in Butcher (1987) Theorem 383B, we see that there is a white-rooted bicolour rooted tree for which the elementary weights of the first and second stages are different. We use Remark 6.6 to construct a C^∞ autonomous, separable Hamiltonian with bounded derivatives for which, for h_1 small and at $t_0 = 0$ and at the origin in phase space, $\mathbf{P}_1 \neq \mathbf{P}_2$ (cf. the proof of Lemma 383A in Butcher (1987)). In a like manner one can construct Hamiltonians that make $\mathbf{Q}_1 \neq \mathbf{Q}_2$, $\mathbf{P}_1 \neq \mathbf{P}_3$, etc. By juxtaposing all the corresponding Hamiltonian systems, we find a Hamiltonian system for which all $\mathbf{P}$-stages are pairwise different and all $\mathbf{Q}$-stages are also pairwise different. □

In the remainder of the subsection, we look at the class $\mathcal{H}^\infty$ of all C^∞, autonomous, separable Hamiltonian functions H defined in the whole of $\mathcal{R}^{2d}$, with bounded partial derivatives of all orders. The corresponding differential systems satisfy a global Lipschitz condition, see Remark 3.1. Hence, given a PRK method (3.24) and $H \in \mathcal{H}^\infty$, there is a constant σ, depending only on the method and on the number of degrees of freedom d, such that, if Λ denotes the lowest upper bound of the modulus of the second partial derivatives of H, then the equations (3.25)–(3.26) that define the stages have a unique solution for

$$h\Lambda < \sigma. \tag{6.20}$$

It follows that, in these circumstances, the mapping $\psi_{h,H}$ associated with the method is a well-defined smooth transformation in $\mathcal{R}^{2d}$. The main result in this section is the following theorem (Abia and Sanz-Serna (1993)) that shows the necessity of (6.9) for symplectiness, even if the Hamiltonian H is restricted to the class $\mathcal{H}^\infty$ and the step size is small.

Theorem 6.6 *If, for a PRK method without equivalent stages, $\psi_{h,H}$ is a canonical transformation for each $H \in \mathcal{H}^\infty$ and each h satisfying (6.20), then (6.9) holds.*

Proof. Our technique of proof is similar to that used by Lasagni (1990) for the case of RK methods and general Hamiltonian problems.

We begin by applying Lemma 6.5 to construct a d-degrees-of-freedom, autonomous separable Hamiltonian $H_0 = T_0 + V_0 \in \mathcal{H}^\infty$ such that, for h small and initial value $(0, 0)$, the stages are pairwise different. Next, for $i = 1, \ldots, s$, let $M_{T,i} = (t_i^{IJ})$ and $M_{V,i} = (v_i^{IJ})$ be $d \times d$ real symmetric matrices with

$$|t_i^{IJ}| \leq 1, \quad |v_i^{IJ}| \leq 1, \qquad I, J = 1, \ldots, d. \tag{6.21}$$

We consider the quadratic Hamiltonian functions

$$\begin{aligned}
H_i(\mathbf{p}, \mathbf{q}) &= T_i(\mathbf{p}) + V_i(\mathbf{q}), \\
T_i(\mathbf{p}) &= T_0 + \frac{\partial T_0}{\partial \mathbf{p}}(\mathbf{p} - \mathbf{P}_i) + \frac{1}{2}(\mathbf{p} - \mathbf{P}_i)^T M_{T,i}(\mathbf{p} - \mathbf{P}_i), \\
V_i(\mathbf{q}) &= V_0 + \frac{\partial V_0}{\partial \mathbf{q}}(\mathbf{q} - \mathbf{Q}_i) + \frac{1}{2}(\mathbf{q} - \mathbf{Q}_i)^T M_{V,i}(\mathbf{q} - \mathbf{Q}_i),
\end{aligned}$$

where T_0, $\partial T_0/\partial \mathbf{p}$ are evaluated at $\mathbf{P}_i$, and V_0, $\partial V_0/\partial \mathbf{q}$ are evaluated at $\mathbf{Q}_i$. Thus, in the neighbourhood of $(\mathbf{P}_i, \mathbf{Q}_i)$, $i = 1, \ldots, s$, H_i and H_0 differ in quadratic terms. Note that H_i depends on h through $\mathbf{P}_i$ and $\mathbf{Q}_i$, but this dependence is not shown in the notation. Finally, let φ be a C^∞ real function defined in $\mathcal{R}^d$ which is $\equiv 1$ in a neighbourhood of the origin and $\equiv 0$ outside a second, larger neighbourhood of the origin, and set, for $0 < \epsilon < 1$,

$$\begin{aligned}
H(\mathbf{p}, \mathbf{q}) = {}& \left(1 - \sum_{i=1}^{s} \varphi\left(\epsilon^{-1}(\mathbf{p} - \mathbf{P}_i)\right)\right) T_0(\mathbf{p}) \\
& + \sum_{i=1}^{s} \varphi\left(\epsilon^{-1}(\mathbf{p} - \mathbf{P}_i)\right) T_i(\mathbf{p}) \\
& + \left(1 - \sum_{i=1}^{s} \varphi\left(\epsilon^{-1}(\mathbf{q} - \mathbf{Q}_i)\right)\right) V_0(\mathbf{q}) \\
& + \sum_{i=1}^{s} \varphi\left(\epsilon^{-1}(\mathbf{q} - \mathbf{Z}_i)\right) V_i(\mathbf{q}). \tag{6.22}
\end{aligned}$$

NECESSITY OF THE SYMPLECTICNESS CONDITIONS

Clearly the function in (6.22) is in $\mathcal{H}^\infty$. Let us now determine, for h small, $\epsilon = \epsilon(h)$ small enough to ensure that $H \equiv H_i$ in a neighbourhood of $(\mathbf{P}_i, \mathbf{Q}_i)$, $i = 1, \ldots, s$; this is possible because the points $\mathbf{P}_i$ are pairwise different and the same happens to the points $\mathbf{Q}_i$. Then, let us reduce the value of h for (6.20) to hold. This is possible because, as $\epsilon \to 0$, the second derivatives of H can be proved to remain bounded. (For instance, when forming the matrix of second derivatives of (6.22), one gets the term

$$\epsilon^{-2} D^2 \varphi \left(\epsilon^{-1}(\mathbf{p} - \mathbf{P}_i) \right) [T_i - T_0];$$

as noted after the construction of T_i, the term in square brackets is of the order of ϵ^2 and this offsets the effect of the factor ϵ^{-2}.)

By assumption, for H in (6.22) and the value of h we have found, the method defines a symplectic transformation. Therefore the second sum in (6.10) vanishes. In order to evaluate this second sum, it is sufficient to observe that the PRK stages for the systems with Hamiltonians H and H_0 are the same, because of the unicity of the solution of the implicit equations guaranteed by (6.20). We can then write

$$0 = \sum_{i,j}(b_i A_{ij} + B_j a_{ji} - b_i B_j) \, d\mathbf{k}_i \wedge d\mathbf{l}_j$$

$$= \sum_{i,j} \sum_{I,J,K} (b_i A_{ij} + B_j a_{ji} - b_i B_j) v_i^{I,J} t_j^{I,K} \, dq_J \wedge dp_K,$$

and the independence of $dq_J \wedge dp_K$ ensures that, for each J and K,

$$\sum_{i,j}(b_i A_{ij} + B_j a_{ji} - b_i B_j) v_i^{I,J} t_j^{I,K} = 0.$$

Since $v_i^{I,J}$ and $t_j^{I,K}$ are arbitrary, subject to (6.21), we have proved that (6.9) must hold. □

6.5.4 Other cases

The necessity of (6.5) for an RK method without equivalent stages to be canonical is implicitly contained in Theorem 6.6; in fact it is enough to see the RK scheme as a particular instance of PRK methods. The case of RKN methods has been addressed by Calvo (1992).

CHAPTER 7

Symplectic order conditions

7.1 Preliminaries

In Chapter 4 we reviewed the standard order conditions that must be imposed on the coefficients of an RK method to ensure that the method has order $\geq r$. Such standard order conditions are *independent* (Butcher (1987) Theorem 306A). This means that if the number of stages s and the elements a_{ij}, b_i of the RK tableau (3.11) are seen as free parameters, then for a given rooted tree $\rho\tau$ the order condition (4.1) is not implied by the order conditions of other rooted trees. When dealing with symplectic RK schemes the coefficients a_{ij}, b_i are constrained by the symplecticness condition (6.5) and hence cannot be seen as free parameters. In fact, Sanz-Serna and Abia (1991) showed that for symplectic RK methods some order conditions are implied by others. Similar considerations apply to PRK (Abia and Sanz-Serna (1993)) and RKN methods (Calvo and Sanz-Serna (1992a)). In the present chapter we review the papers just mentioned and explain how to write the order conditions for symplectic RK, PRK or RKN methods.

It is advisable to comment first on some graph-theoretic facts that play an important role later. Let us return to Fig. 4.1 that displays the rooted trees of order ≤ 4. If in $\rho\tau_{3,1}$ and $\rho\tau_{3,2}$ we disregard the locations of the crosses that highlight the roots, then both graphs are identical; they consist of the same vertices joined by the same set of edges. The graph obtained by disregarding the location of the root in a rooted tree is called a *free tree* or, simply, a *tree*. Thus a tree τ can be seen as an equivalence class of rooted trees. On the right of Fig. 4.1 we have displayed the trees of order ≤ 4; a tree and the rooted trees associated with it (or belonging to it) appear in the same row.

Similar considerations apply to Fig. 4.3. There bicolour rooted trees $\beta\rho\tau$ give rise to *bicolour trees* $\beta\tau$. Each $\beta\tau$ can be seen as an equivalence class of bicolour rooted trees.

It is useful to compare Figs. 4.1 and 4.3. Each rooted tree $\rho\tau$ gives rise to *two* bicolour rooted trees $\beta\rho\tau$: the root can be coloured either white or black and, once the colour of the root has been chosen, the colour of the remaining vertices is determined by the rule that a father and its sons have different colours. In particular, there are 8 rooted trees in Fig. 4.1 and 16 bicolour rooted trees in Fig. 4.3.

Things get slightly more complicated when it comes to trees and bicolour trees. There are 5 trees in Fig. 4.1 and only 8 in Fig. 4.3, rather than the 10 one might have expected. Why? Most trees can be coloured in two different ways, e.g. $\tau_{3,1}$ can be coloured with a white central vertex to give rise to $\beta\tau_{3,1}$ or with a black central vertex to originate $\beta\tau_{3,2}$. Nevertheless, some trees can be coloured in only one way and therefore correspond to only one bicolour tree. For instance, $\tau_{2,1}$ can be coloured in *only one way* to give rise to $\beta\tau_{2,1}$; if the colours of the picture of $\beta\tau_{2,1}$ in Fig. 4.3 are reversed then we obtain a different picture of *the same* bicolour tree $\beta\tau_{2,1}$, rather than a new bicolour tree.

Those trees that can be coloured in only one way were called *superfluous* by Sanz-Serna and Abia (1991). It is easy to see that superfluous trees must have an even number of vertices. Furthermore a superfluous tree of order $2r$ can be obtained by taking two copies of a rooted tree of order r and joining them by their roots. For instance the superfluous $\tau_{4,1}$ is generated by taking two copies of $\rho\tau_{2,1}$ and joining them through the two higlighted vertices.

Finally, in Fig. 4.4, special Nyström rooted trees $\sigma\nu\rho\tau$ group themselves into *special Nyström trees* $\sigma\nu\tau$.

7.2 Order of symplectic Runge-Kutta methods

We begin with a preliminary result. Let us consider a tree τ of order (number of vertices) $r \geq 2$ and let us label its vertices with letters $i, j, k, \ldots$ For any pair of adjacent vertices, labelled, say, i and j, we consider four rooted trees as follows. Denote by $\rho\tau_i$ (respectively, $\rho\tau_j$) the rooted tree that is obtained from τ by choosing the vertex i (respectively, j) to play the role of root (see Fig. 7.1). Furthermore, we denote by $\rho\tau_I$ and $\rho\tau_J$ the rooted trees, with roots at i and j respectively, that arise when the edge joining i to j is removed from τ.

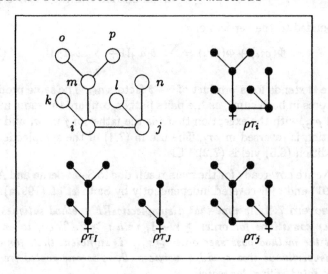

Figure 7.1. *Graphs for Lemma 7.1*

Lemma 7.1 *With the above notation,*

$$\frac{1}{\gamma(\rho\tau_i)} + \frac{1}{\gamma(\rho\tau_j)} = \frac{1}{\gamma(\rho\tau_I)}\frac{1}{\gamma(\rho\tau_J)}, \qquad (7.1)$$

and, for the elementary weights of a symplectic RK method,

$$\Phi(\rho\tau_i) + \Phi(\rho\tau_j) = \Phi(\rho\tau_I)\Phi(\rho\tau_J). \qquad (7.2)$$

Hence, for a symplectic RK method of order $\geq r-1$,

$$\Phi(\rho\tau_i) + \Phi(\rho\tau_j) = \frac{1}{\gamma(\rho\tau_i)} + \frac{1}{\gamma(\rho\tau_j)}, \qquad (7.3)$$

so that the order condition for $\rho\tau_i$ holds if and only if the order condition for $\rho\tau_j$ holds.

Proof. From the recursive definition of γ (see Section 4.1)

$$\gamma(\rho\tau_i) = r\gamma(\rho\tau_J)\left[\frac{\gamma(\rho\tau_I)}{r(\rho\tau_I)}\right],$$

$$\gamma(\rho\tau_j) = r\gamma(\rho\tau_I)\left[\frac{\gamma(\rho\tau_J)}{r(\rho\tau_J)}\right],$$

where $r(\rho\tau_I)$ and $r(\rho\tau_J)$ are the orders of $\rho\tau_I$ and $\rho\tau_J$. Then (7.1) is a direct consequence of the equality $r(\rho\tau_I) + r(\rho\tau_J) = r$.

The left-hand side of (7.2) can be written (see Section 4.1) in terms of two sums whose summation indices are the labels we appended to the vertices of τ:

$$\Phi(\rho\tau_i) + \Phi(\rho\tau_j) = \sum_{i,j,\ldots} b_i a_{ij} \Pi + \sum_{i,j,\ldots} b_j a_{ij} \Pi. \qquad (7.4)$$

Here Π stands for a product of $r-2$ factors a_{kl}. The same product features in both sums, as the pairs [father, son] are the same in $\rho\tau_i$ and $\rho\tau_j$, with the exception that i is the father of j in $\rho\tau_i$ and this relation is reversed in $\rho\tau_j$. The use in (7.4) of the symplecticness condition (6.5) yields (7.2). □

We are now ready for the main result due to Sanz-Serna and Abia (1991) and rediscovered independently by Saito *et al.* (1992a).

Theorem 7.2 *Assume that a symplectic RK method satisfies the order conditions for order $\geq r - 1$, with $r \geq 2$. Then, to ensure that the method possesses order $\geq r$, it is sufficient that, for each nonsuperfluous tree τ with r vertices, there is one rooted tree $\rho\tau$ associated with τ for which*

$$\Phi(\rho\tau) = \frac{1}{\gamma(\rho\tau)}. \qquad (7.5)$$

Proof. We have to check that, under the hypotheses of the theorem, (7.5) actually holds for each rooted tree $\rho\tau$ of order r.

Choose first a nonsuperfluous tree τ. Then, by hypothesis, (7.5) holds for a suitable rooted tree $\rho\tau_i$ belonging to τ. We now consider the construction in the lemma, with j any of the vertices adjacent to i. By (7.3) the order condition (7.5) corresponding to $\rho\tau_j$ is also satisfied. Since any two vertices of a tree can be joined through a chain of pairwise adjacent vertices, the iteration of this argument leads to the conclusion that the method satisfies the order conditions that arise from any rooted tree in τ.

In the case of a superfluous tree τ, we observe that, by definition of a superfluous tree, it is possible to choose adjacent vertices i and j such that $\rho\tau_i$ and $\rho\tau_j$ are in fact the same rooted tree. Then (7.3) shows that (7.5) holds for the rooted tree $\rho\tau_i$. After this, we proceed as above to inductively show that (7.5) holds for *all* rooted trees in τ. □

Example 7.1 For $r = 2$ there is only one tree $\tau_{2,1}$. This is superfluous and hence the sufficient condition of the theorem is empty: any consistent, symplectic RK method has actually order ≥ 2. □

Example 7.2 For $r = 3$ there is again only one tree $\tau_{3,1}$. This comprises two rooted trees $\rho\tau_{3,1}$ and $\rho\tau_{3,2}$. Hence for a consis-

ORDER OF SYMPLECTIC PARTITIONED METHODS

Table 7.1. *Number of order conditions for RK methods*

Order	General RK	Symplectic RK
1	1	1
2	2	1
3	4	2
4	8	3
5	17	6
6	37	10
7	85	21
8	200	40

tent, symplectic RK method to have order ≥ 3 it is enough to impose either the order condition (4.4) *or* the order condition (4.5). The two order equations for $r = 3$, that are independent for general RK methods, have become mutually equivalent for symplectic methods. □

Example 7.3 For $r = 4$ there is only one nonsuperfluous tree $\tau_{4,2}$. For a symplectic RK method of order ≥ 3 to have order ≥ 4, we impose *either* the order condition for $\rho\tau_{4,3}$ *or* the order condition for $\rho\tau_{4,4}$. □

In general, we see that for symplectic methods there is one order equation per nonsuperfluous tree rather than one order equation per rooted tree. The reduction in the number of order conditions is borne out in Table 7.1.

7.3 Order of symplectic Partitioned methods

We now turn our attention to PRK methods. We have the following result (Abia and Sanz-Serna (1993)), whose proof is almost identical to that of Theorem 7.2 and will not be given.

Theorem 7.3 *Assume that a symplectic PRK method satisfies the order conditions for order $\geq r - 1$, with $r \geq 2$. Then, to ensure that the method possesses order $\geq r$, it is sufficient that, for each bicolour tree $\beta\tau$ with r vertices, there is one bicolour rooted tree $\beta\rho\tau$ associated with $\beta\tau$ for which*

$$\Phi(\beta\rho\tau) = \frac{1}{\gamma(\beta\rho\tau)}. \tag{7.6}$$

Table 7.2. *Number of order conditions for PRK and RKN methods*

Order	General PRK	Symplectic PRK	General RKN	Symplectic RKN
1	2	2	1	1
2	4	3	2	2
3	8	5	4	4
4	16	8	7	6
5	34	14	13	10
6	74	24	23	15
7	170	46	43	25
8	400	88	79	39

Let us present a few instances of the application of the theorem.

Example 7.4 For $r = 2$ there is only one bicolour tree $\beta\tau_{2,1}$ (see Fig. 4.3). This comprises two bicolour rooted trees $\beta\rho\tau_{2,1,w}$ and $\beta\rho\tau_{2,1,b}$. For a consistent, symplectic PRK method to have order ≥ 2, we therefore impose either

$$\sum_{ij} b_i A_{ij} = \frac{1}{2} \qquad (7.7)$$

or

$$\sum_{ij} B_i a_{ij} = \frac{1}{2}. \quad \Box \qquad (7.8)$$

Example 7.5 For $r = 3$ there are two bicolour trees $\beta\tau_{3,1}$ and $\beta\tau_{3,2}$. Each comprises two bicolour rooted trees. Thus for a symplectic PRK method of order ≥ 2 to have order ≥ 3, we impose one of the two following conditions

$$\sum_{ijk} B_i a_{ij} A_{jk} = \frac{1}{6}, \qquad \sum_{ijk} b_i A_{ij} A_{ik} = \frac{1}{3} \qquad (7.9)$$

(associated respectively with $\beta\rho\tau_{3,1,b}$ and $\beta\rho\tau_{3,2,w}$) *and* one of the two conditions

$$\sum_{ijk} b_i A_{ij} a_{jk} = \frac{1}{6}, \qquad \sum_{ijk} B_i a_{ij} a_{ik} = \frac{1}{3}$$

(associated respectively with $\beta\rho\tau_{3,1,w}$ and $\beta\rho\tau_{3,2,b}$). $\Box$

Table 7.2 shows the reduction in the number of order conditions introduced by symplecticness.

ORDER OF SYMPLECTIC RUNGE-KUTTA-NYSTRÖM METHODS

Remark 7.1 It is perhaps useful to emphasize that in the RK case superfluous trees do not bring in an order condition, while for PRK schemes bicolour trees resulting from colouring a superfluous tree do introduce an order condition. However note that to a nonsuperfluous tree τ there correspond two bicolour trees $\beta\tau$ and therefore two symplectic PRK order conditions; to a superfluous tree there corresponds only one bicolour tree and one symplectic PRK order condition. □

7.4 Order of symplectic Runge-Kutta-Nyström methods

Let us now consider the situation for symplectic Runge-Kutta-Nyström methods. The role played in the RK case by the construction in Fig. 7.1 is now played by the construction in Fig. 7.2. Now i and j are labels attached to fat vertices joined to each other by a meagre vertex v. The special Nyström rooted trees $\sigma\nu\rho\tau_i$ and $\sigma\nu\rho\tau_j$ correspond to the whole underlying special Nyström tree with the root at i or j, respectively. The special Nyström rooted trees $\sigma\nu\rho\tau_I$ and $\sigma\nu\rho\tau_J$ arise after suppression of the vertex v and the edges that join v to i and j. Finally the special Nyström rooted

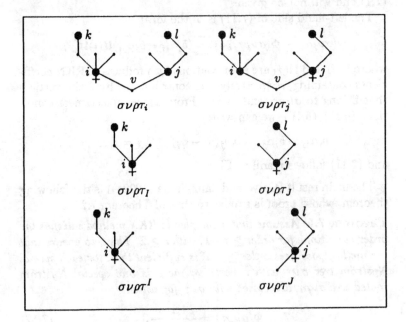

Figure 7.2. *Graphs for Lemma 7.4*

trees $\sigma\nu\rho\tau^I$ and $\sigma\nu\rho\tau^J$ are the result of appending v to the root of $\sigma\nu\rho\tau_I$ and $\sigma\nu\rho\tau_J$, respectively.

Lemma 7.4 *With the above notation,*

$$\frac{1}{\gamma(\sigma\nu\rho\tau_i)} - \frac{1}{\gamma(\sigma\nu\rho\tau_j)}$$
$$= \frac{1}{\gamma(\sigma\nu\rho\tau^I)}\frac{1}{\gamma(\sigma\nu\rho\tau_J)} - \frac{1}{\gamma(\sigma\nu\rho\tau_I)}\frac{1}{\gamma(\sigma\nu\rho\tau^J)} \quad (7.10)$$

and, for the elementary weights of a symplectic RKN method,

$$\Phi(\sigma\nu\rho\tau_i) - \Phi(\sigma\nu\rho\tau_j)$$
$$= \Phi(\sigma\nu\rho\tau^I)\Phi(\sigma\nu\rho\tau_J) - \Phi(\sigma\nu\rho\tau_I)\Phi(\sigma\nu\rho\tau^J). \quad (7.11)$$

Hence, for a symplectic RK method of order $\geq r-1$,

$$\Phi(\sigma\nu\rho\tau_i) - \Phi(\sigma\nu\rho\tau_j) = \frac{1}{\gamma(\sigma\nu\rho\tau_i)} - \frac{1}{\gamma(\sigma\nu\rho\tau_j)},$$

so that the order condition for $\sigma\nu\rho\tau_i$ *holds if and only if the order condition for* $\sigma\nu\rho\tau_j$ *holds.*

Proof. The argument to derive (7.10) is similar to that used for (7.1) and will not be given.

The left-hand side of (7.11) is of the form

$$\Phi(\sigma\nu\rho\tau_i) - \Phi(\sigma\nu\rho\tau_j) = \sum(b_i\alpha_{ij} - b_j\alpha_{ji})\Pi(i)\Pi(j),$$

where $\Pi(i)$ and $\Pi(j)$ are abbreviations for products of RKN coefficients containing, respectively, the contributions from the vertices that belong to $\sigma\nu\rho\tau_I$ and $\sigma\nu\rho\tau_J$. From the symplecticness conditions (6.15)–(6.16), we can write

$$b_i\alpha_{ij} - b_j\alpha_{ji} = b_ib_j(\gamma_i - \gamma_j), \quad i,j = 1,\ldots,s$$

and (7.11) follows readily. □

The main result (Calvo and Sanz-Serna (1992a)) is the following theorem, whose proof is similar to that of Theorem 7.2.

Theorem 7.5 *Assume that a symplectic RKN method satisfies the order conditions for order* $\geq r-1$, *with* $r \geq 2$. *Then, to ensure that the method possesses order* $\geq r$, *it is sufficient that, for each special Nyström tree* $\sigma\nu\tau$ *with* r *vertices, there is one special Nyström rooted tree* $\sigma\nu\rho\tau$ *associated with* $\sigma\nu\tau$ *for which*

$$\Phi(\sigma\nu\rho\tau) = \frac{1}{\gamma(\sigma\nu\rho\tau)}. \quad (7.12)$$

7.5 Homogeneous form of the order conditions

7.5.1 Motivation

Let us return to the RK case. We noticed above that, for $r = 3$ and symplectic RK methods it is enough to impose the order condition associated with $\rho\tau_{3,1}$ or the order condition associated with $\rho\tau_{3,2}$. Thus, when writing the order conditions for symplectic RK methods, some rooted trees are taken into account while some are left out. There is an alternative form of the order conditions in which all rooted trees play a symmetric role. This is the so-called *homogeneous form* of the order conditions, introduced by Sanz-Serna and Abia (1991). The full meaning of the homogeneous form will be clear when in Chapter 11 we study the canonical theory of the order.

7.5.2 The Partitioned Runge-Kutta case

We present first the homogeneous form of the order conditions for PRK methods (Abia and Sanz-Serna (1993)).

Theorem 7.6 *Assume that a symplectic PRK method satisfies the order conditions for order $\geq r - 1$, with $r \geq 2$. Then, to ensure that the method possesses order $\geq r$, it is necessary and sufficient that, for each bicolour tree $\beta\tau$ with r vertices*

$$\sum_{\substack{\beta\rho\tau \in \beta\tau \\ \beta\rho\tau \text{ has white root}}} \alpha(\beta\rho\tau)\gamma(\beta\rho\tau)\Phi(\beta\rho\tau)$$

$$= \sum_{\substack{\beta\rho\tau \in \beta\tau \\ \beta\rho\tau \text{ has black root}}} \alpha(\beta\rho\tau)\gamma(\beta\rho\tau)\Phi(\beta\rho\tau). \quad (7.13)$$

Proof. The necessity is considered first. If the method has order r, then (7.13) reads

$$\sum_{\substack{\beta\rho\tau \in \beta\tau \\ \beta\rho\tau \text{ h.w.r.}}} \alpha(\beta\rho\tau) = \sum_{\substack{\beta\rho\tau \in \beta\tau \\ \beta\rho\tau \text{ h.b.r.}}} \alpha(\beta\rho\tau). \quad (7.14)$$

If the tree τ corresponding to $\beta\tau$ is superfluous, then the last equality is true: if a bicolour rooted tree contributes to the sum in the right-hand side, then its photographic negative (with white and black vertices interchanged) contributes to the sum in the left with exactly the same value of α. For the nonsuperfluous case, the graph-theoretic result (7.14) is proved in Sanz-Serna and Abia (1991), Theorem 4.5.

Let us show that (7.13) is also sufficient for the method to have order r. We begin by noticing that, if in a situation similar to that in Lemma 7.1, we consider two bicolour rooted trees $\beta\rho\tau_i$, $\beta\rho\tau_j$ belonging to the same bicolour rooted tree $\beta\tau$ and with roots in adjacent vertices i and j then, as in (7.3),

$$\Phi(\beta\rho\tau_i) + \Phi(\beta\rho\tau_j) = \frac{1}{\gamma(\beta\rho\tau_i)} + \frac{1}{\gamma(\beta\rho\tau_j)}. \qquad (7.15)$$

If the method did not have order r, then there is at least a bicolour rooted tree, say $\beta\rho\tau_i$ whose order condition is not satisfied. To be definite, let us assume that this has a white root and that

$$\gamma(\beta\rho\tau_i)\Phi(\beta\rho\tau_i) > 1. \qquad (7.16)$$

We now consider a second bicolour rooted tree $\beta\rho\tau_j$, belonging to the same bicolour tree as $\beta\rho\tau_i$ and with root adjacent to that of $\beta\rho\tau_i$. Combining (7.15) and (7.16) we see that for $\beta\rho\tau_j$, whose root is black, the product $\gamma\Phi$ is < 1. The iteration of this argument shows that $\gamma\Phi$ is > 1 for bicolour rooted trees with white roots and < 1 for bicolour rooted trees with black roots. When this is taken into (7.14) we obtain a contradiction with (7.13). $\square$

Example 7.6 For $\beta\tau_{2,1}$ we noticed above that, if the standard form is used, we must consider either (7.7) or (7.8). When using the homogeneous form, we simply write

$$2\sum_{ij} b_i A_{ij} = 2\sum_{ij} B_i a_{ij}. \quad \square \qquad (7.17)$$

Example 7.7 For $\beta\tau_{3,1}$ we noticed above that, if the standard form is used, we must consider one of the two equations in (7.9). When using the homogeneous form, we simply write

$$3\sum_{ijk} b_i A_{ij} A_{ik} = 6\sum_{ijk} B_i a_{ij} A_{jk}. \qquad (7.18)$$

Similarly for $\beta\tau_{3,2}$ the homogeneous form reads

$$6\sum_{ijk} b_i A_{ij} a_{jk} = 3\sum_{ijk} B_i a_{ij} a_{ik}. \quad \square \qquad (7.19)$$

7.5.3 Other cases

For symplectic RKN methods a homogeneous form of the order equations is available (Calvo (1992)), but rather messy. The interested reader is referred to the original work by Calvo.

For the RK case, a derivation of the order conditions can be seen in Sanz-Serna and Abia (1991). In the context of the present book, the best way to proceed is to note that, as pointed out in Remark 4.2, it is possible to obtain the RK order conditions by seeing an RK method as a particular instance of PRK methods, where both tableaux just happen to coincide. It is therefore feasible to write the homogeneous form of the order conditions for RK methods by applying (7.13) with $A_{ij} = a_{ij}$ and $B_i = b_i$. Let us do this.

Example 7.8 For $r = 2$, we must consider (7.17), which now, with small and capital letters coinciding, becomes an identity. Hence, there is no homogenous order condition for a consistent, symplectic RK method to have order ≥ 2. This of course agrees with our earlier findings in Example 7.1 when using the standard form of the order conditions. □

Example 7.9 For $r = 3$ we have to consider the conditions (7.18) and (7.19). When small and capital coefficients are the same, these two conditions coincide and therefore there is only one homogeneous order condition for symplectic RK methods, namely

$$3\sum_{ijk} b_i a_{ij} a_{ik} = 6 \sum_{ijk} b_i a_{ij} a_{jk}.$$

This is to be compared with (4.4)–(4.5). □

The general trend should now be clear. For a superfluous tree, there is only one homogeneous PRK order condition and this becomes an identity if the PRK method is actually an RK method. This matches the fact that only nonsuperfluous trees are considered in Theorem 7.2. On the other hand, a nonsuperfluous tree gives rise to two bicolour trees, one the photographic negative of the other. The homogeneous PRK order conditions corresponding to this pair of bicolour trees are obtained from one another by interchanging capital and small coefficients. Hence they become identical if the PRK method reduces to an RK method. This yields one homogeneous RK condition per nonsuperfluous tree (cf. Theorem 7.2).

CHAPTER 8

Available symplectic methods

8.1 Symplecticness of the Gauss methods

In this chapter we present examples of symplectic RK, PRK and RKN methods. The emphasis is on low-order, $r \leq 4$, methods. High-order methods are presented in Chapter 13.

We begin with the Gauss methods of Subsection 3.3.2. The following result was first proved by Sanz-Serna (1988) by using the characterization of symplectic RK methods in (6.5). The direct proof we present here has been communicated to us by E. Hairer and G. Wanner.

Theorem 8.1 *For each $s = 1, 2, \ldots$ the s-stage Gauss method is symplectic.*

Proof. Let us respectively denote by $\mathbf{p}(t)$ and $\mathbf{q}(t)$ the collocation polynomials corresponding to the variables $\mathbf{p}$ and $\mathbf{q}$ (see Subsection 3.3.2). Since $\mathbf{p}^{n+1} = \mathbf{p}(t_{n+1})$, $\mathbf{p}^n = \mathbf{p}(t_n)$ and similarly for $\mathbf{q}$, we may write

$$d\mathbf{p}^{n+1} \wedge \mathbf{q}^{n+1} - d\mathbf{p}^n \wedge \mathbf{q}^n = \int_{t_n}^{t_{n+1}} \frac{d}{dt}(d\mathbf{p} \wedge d\mathbf{q})\, dt.$$

Now $\mathbf{p}(t)$ and $\mathbf{q}(t)$ are polynomials of degree $\leq s$ in t, so that the integrand is of degree $\leq 2s - 1$. It follows that it can be integrated exactly by the Gauss quadrature rule associated with the Gauss RK method (cf. (3.17)). Thus

$$d\mathbf{p}^{n+1} \wedge \mathbf{q}^{n+1} - d\mathbf{p}^n \wedge d\mathbf{q}^n = h_{n+1} \sum_{i=1}^{s} \frac{d}{dt}(d\mathbf{p} \wedge d\mathbf{q})\big|_{t_n + c_i h_{n+1}}.$$

With the notation used in the proof of Theorem 6.1, the formulae (3.15)–(3.16), that relate the intermediate stages with the collocation polynomial, read

$$\mathbf{p}(t_n + c_i h_{n+1}) = \mathbf{P}_i,$$

$$\mathbf{q}(t_n + c_i h_{n+1}) = \mathbf{Q}_i,$$
$$\dot{\mathbf{p}}(t_n + c_i h_{n+1}) = \mathbf{k}_i,$$
$$\dot{\mathbf{q}}(t_n + c_i h_{n+1}) = \mathbf{l}_i.$$

Hence

$$d\mathbf{p}^{n+1} \wedge d\mathbf{q}^{n+1} - d\mathbf{p}^n \wedge \mathbf{q}^n = h_{n+1} \sum_{i=1}^{s} (d\mathbf{k}_i \wedge d\mathbf{Q}_i + d\mathbf{P}_i \wedge d\mathbf{l}_i).$$

In the proof of Theorem 6.1 it was established that the term in brackets vanishes for each i; symplecticness follows. □

The Gauss methods are not only symplectic; they have good stability properties when applied to stiff systems, namely they are *B-stable* and hence *A-stable* (Dekker and Verwer (1984)).

The two-stage, order-4 method has been successfully tested by Pullin and Saffman (1991) in a Hamiltonian problem arising in fluid mechanics.

8.2 Diagonally implicit Runge-Kutta methods

8.2.1 General format

We recall that symplectic RK methods cannot be explicit. For nonstiff Hamiltonian problems, the Gauss methods offer the combined advantages of symplecticness, possibility of high order and easy implementation with functional iteration (see Chapter 5). For *stiff* Hamiltonian problems, the Gauss methods present the advantage of good stability properties. However, Newton iteration has to be used and the cost of the required linear algebra may be high. It is then of interest to investigate the existence of diagonally implicit RK methods.

When looking for symplectic RK methods (diagonally implicit or not), it may be assumed that all the weights b_i are $\neq 0$. In fact, if $b_j = 0$, then the symplecticness condition (6.5) implies that $b_i a_{ij} = 0$ for all i and therefore the j-th stage, which is not directly contributing to the final approximations (6.3)–(6.4), is not contributing either to any other stage with nontrivial b_i: thus the method is equivalent to a method with fewer stages.

Under the assumption $b_i \neq 0$, it is straightforward to check that diagonally implicit RK schemes satisfy the symplecticness condition (6.5) if and only if they have the tableau

$$\begin{array}{c|ccccc} b_1/2 & 0 & 0 & \cdots & 0 \\ b_1 & b_2/2 & 0 & \cdots & 0 \\ b_1 & b_2 & b_3/2 & \cdots & 0 \\ \vdots & \vdots & \vdots & \ddots & \vdots \\ b_1 & b_2 & b_3 & \cdots & b_s/2 \\ \hline & b_1 & b_2 & b_3 & \cdots & b_s \end{array} \qquad (8.1)$$

A step of length h with the method (8.1) is just a concatenation (see Section 3.6) of an implicit midpoint step of length $b_1 h$, followed by an implicit midpoint step of length $b_2 h$, etc... Hence the symplecticness of the method with tableau (8.1) may be established, without consideration of (6.5), by noticing that the midpoint rule is symplectic and that the composition of symplectic mappings is a symplectic mapping.

Another, more practical, consequence of the concatenation nature of (8.1) is that it implies storage savings. Once the vector

$$\mathbf{y}^{n+b_1} = \mathbf{y}^n + h_{n+1} b_1 \mathbf{F}(\mathbf{Y}_1, t_n + b_1 h_{n+1})$$

resulting from the first midpoint substep is known, there is no need to keep $\mathbf{y}^n$ in storage. The vector $\mathbf{y}^{n+b_1}$ is the starting point for a new midpoint subset and can be overwritten on $\mathbf{y}^n$. The result $\mathbf{y}^{n+b_1+b_2}$ after two substeps is overwritten on $\mathbf{y}^{n+b_1}$ etc.

Furthermore there are also small arithmetic savings with respect to general diagonally implicit RK methods. In the general case the j-th evaluation $\mathbf{F}(\mathbf{Y}_j, t_n + c_j h_{n+1})$, before contributing to the i-th stage, is multiplied by constants a_{ij}, $i = j, \ldots, s$ that *vary* with i. Here the contribution of the j evaluation to each of the subsequent stages does not depend on i and is computed once and for all when forming $\mathbf{y}^{n+b_1+\ldots+b_j}$.

The *adjoint* of the method (8.1) is the diagonally implicit symplectic RK method with weights $\bar{b}_i = b_{s-i+1}$, $i = 1, \ldots, s$. In particular (8.1) is symmetric if $b_1 = b_s$, $b_2 = b_{s-1}$, ...

8.2.2 Specific methods

For any choice of the free parameters b_i, subject to consistency

$$\sum_{i=1}^{s} b_i = 1, \qquad (8.2)$$

the method (8.1) has, as any other consistent symplectic RK procedure, order ≥ 2.

To achieve order ≥ 3, there is one extra order condition (see Table 7.1). It is convenient to use the homogeneous form. This is given by (7.18) with capital letters changed into lower case letters, since we are now in the RK case. After some manipulation, the condition reads

$$\sum_{i=1}^{s} b_i^3 = 0. \qquad (8.3)$$

Clearly with $s = 2$, the system (8.2)–(8.3) has no real solutions. The minimum number of stages is then three, with a one-parameter family of order-3 methods. Sanz-Serna and Abia (1991) choose the member of the family determined by $b_1 = b_3$. This ensures (see the preceding subsection) the symmetry of the method and hence order 4 (see Section 3.6). The method has coefficients

$$b_1 = b_3 = \frac{1}{3}(2 + 2^{1/3} + 2^{-1/3}), \quad b_2 = 1 - 2b_3, \qquad (8.4)$$

and has been studied by Frutos and Sanz-Serna (1992).

8.3 Other symplectic Runge-Kutta methods

Cooper (1987) appears to have been the first in investigating RK methods that satisfy the symplecticness condition (6.5). However Cooper was not concerned with Hamiltonian problems.

Iserles (1991) (see also Iserles and Nørsett (1991)) studies symplectic RK methods for which the matrix $A = (a_{ij})$ of the coefficients has only real eigenvalues, a property that may lead to cheaper linear algebra, see Remark 5.2.

A symplectic RK method with three stages and order 4 is constructed by Maeda (1991). This method has positive weights and hence it is *B-stable*.

Saito *et al.* (1992b) point out that if b_i and c_i, $i = 1, \ldots, s$ are the weights and abscissae of a quadrature formula exact for polynomials of degree ≤ 1 (i.e., $\sum b_i = 1$, $\sum_i b_i c_i = 1/2$), then the family of symplectic RK methods with weights b_i and abscissae c_i depends linearly on $(s-1)(s-2)/2$ parameters.

A systematic study of symplectic RK methods of high order is carried out in Sun (1992a) by using the W-transformation of Hairer and Wanner (1981) (see also Hairer and Wanner (1991), Chapter IV). Amongst known families of RK methods, not only the Gauss methods, but also the Lobatto III E (Nørsett and Wanner (1981)) and Lobatto III S (Chan (1990)) are symplectic. Sun introduces two new families of RK methods, Radau I B and Radau II B, whose members are symplectic.

8.4 Explicit Partitioned Runge-Kutta methods

8.4.1 General format

Symplectic PRK methods are only applicable to *separable* Hamiltonian systems. In this respect the class of symplectic PRK methods seems less appealing than the class of symplectic RK methods. However the good news is that there are PRK methods that are *explicit*. In fact, we noticed in Section 3.4 that a PRK method with $a_{ij} = 0$ for $i \leq j$ and $A_{ij} = 0$ for $i < j$ is effectively explicit. When these conditions on the a_{ij}, A_{ij} are taken into the symplecticness requirement (6.9), the following family of explicit, symplectic methods arises:

$$\begin{array}{|ccccc|ccccc|}
0 & 0 & 0 & \cdots & 0 & B_1 & 0 & 0 & \cdots & 0 \\
b_1 & 0 & 0 & \cdots & 0 & B_1 & B_2 & 0 & \cdots & 0 \\
b_1 & b_2 & 0 & \cdots & 0 & B_1 & B_2 & B_3 & \cdots & 0 \\
\vdots & \vdots & \vdots & \ddots & \vdots & \vdots & \vdots & \vdots & \ddots & \vdots \\
b_1 & b_2 & b_3 & \cdots & 0 & B_1 & B_2 & B_3 & \cdots & B_s \\
\hline
b_1 & b_2 & b_3 & \cdots & b_s & B_1 & B_2 & B_3 & \cdots & B_s
\end{array} \qquad (8.5)$$

Remark 8.1 Okunbor and Skeel (1992b) have proved that any explicit, symplectic PRK method can be rewritten in the form (8.5), possibly after renumbering the stages, adding dummy stages or suppressing redundant stages. These manipulations are illustrated below in some specific instances. □

Note that with s stages there are $2s$ free parameters in (8.5) and that the method (8.5) can be denoted in the shorter form

$$(b_1, b_2, \ldots, b_s)[B_1, B_2, \ldots, B_s]. \qquad (8.6)$$

In this notation, the square brackets enclose the weights of the tableau in (8.5) with a nontrivial main diagonal; the weights on the left are used to advance the p variables, those on the right the q variables.

Methods of the form (8.5) possess the favourable property that they may be implemented while only storing two d-dimensional vectors: $\mathbf{P}_1$ is nothing but $\mathbf{p}^n$, $\mathbf{Q}_1$ can be overwritten on $\mathbf{q}^n$, $\mathbf{P}_2$ can be overwritten on $\mathbf{P}_1$, etc... The computation proceeds in the following form:

$$Q_0 = q^n,$$
$$P_1 = p^n,$$
for $i = 1, \ldots, s$
$$Q_i = Q_{i-1} + h_{n+1} B_i g(P_i),$$
$$P_{i+1} = P_i + h_{n+1} b_i f(Q_i, t_n + C_i h_{n+1}),$$
$$q^{n+1} = Q_s,$$
$$p^{n+1} = P_{s+1}.$$

Furthermore, and as pointed out with dealing with diagonally implicit, symplectic RK schemes, the fact that the tableaux in (8.5) are constant along columns implies some minor arithmetic savings.

8.4.2 An alternative format

Due to the mathematical symmetry of the p and q variables in a Hamiltonian problem, the tableaux

$$\begin{array}{|ccccc|ccccc|} b_1 & 0 & 0 & \cdots & 0 & 0 & 0 & 0 & \cdots & 0 \\ b_1 & b_2 & 0 & \cdots & 0 & B_1 & 0 & 0 & \cdots & 0 \\ b_1 & b_2 & b_3 & \cdots & 0 & B_1 & B_2 & 0 & \cdots & 0 \\ \vdots & \vdots & \vdots & \ddots & \vdots & \vdots & \vdots & \vdots & \ddots & \vdots \\ b_1 & b_2 & b_3 & \cdots & b_s & B_1 & B_2 & B_3 & \cdots & 0 \\ \hline b_1 & b_2 & b_3 & \cdots & b_s & B_1 & B_2 & B_3 & \cdots & B_s \end{array} \quad (8.7)$$

also correspond to explicit symplectic PRK algorithms. Note that now the tableau with a trivial main diagonal now corresponds to the q variables. We employ the notation

$$[b_1, b_2, \ldots, b_s](B_1, B_2, \ldots, B_s). \qquad (8.8)$$

This should be compared with (8.6); the locations of square and curved brackets have been interchanged.

The implementation now is

$$P_0 = p^n,$$
$$Q_1 = q^n,$$
for $i = 1, \ldots, s$
$$P_i = P_{i-1} + h_{n+1} b_i f(Q_i, t_n + C_i h_{n+1}),$$
$$Q_{i+1} = Q_i + h_{n+1} B_i g(P_i),$$

$$\mathbf{p}^{n+1} = \mathbf{P}_s,$$
$$\mathbf{q}^{n+1} = \mathbf{Q}_{s+1}.$$

Remark 8.2 The existence of the alternative format (8.7) does not contradict the result by Okunbor and Skeel reported in Remark 8.1. In fact, the method (8.7) can be recast in the format (8.5) by adding a dummy stage; more precisely (8.8) can be rewritten as

$$(b_1, b_2, \ldots, b_s, 0)[0, B_1, B_2, \ldots, B_s].$$

In a like manner methods in the format (8.5) can be rewritten in the format (8.7) with the addition of an extra (dummy) stage. Hence only one of the formats is strictly necessary from a mathematical point of view. However, if possible, it is best to write the procedures without dummy stages and we shall use both formats. □

The adjoint of the method (8.6), computed by the rule in Section 3.6, turns out to be

$$[b_s, \ldots, b_1](B_s, \ldots, B_1).$$

Similarly, the adjoint of (8.8) is

$$(b_s, \ldots, b_1)[B_s, \ldots, B_1].$$

8.4.3 Specific methods: orders 1 and 2

For order 1, there are two order conditions (see Table 7.2). Since (8.5) has $2s$ free parameters, we can choose $s = 1$, leading to the method

$$(1)[1].$$

In the alternative format (8.7) we find the method

$$[1](1);$$

these first-order methods are mutually adjoint.

For order 2 there are in all three order conditions, so that $s = 2$ leaves room for a free parameter. In the format (8.5) we can choose $b_2 = 0$; this saves the evaluation of $\mathbf{f}$ at $\mathbf{Q}_2$. The method is

$$(1, 0)[1/2, 1/2]. \tag{8.9}$$

Note that the choice $b_2 = 0$ also implies an FSAL property (see Remark 5.1), namely $\mathbf{P}_2$ equals $\mathbf{p}^{n+1}$, which is the first p stage at the next time level. Hence not only $\mathbf{f}$ but also $\mathbf{g}$ is essentially evaluated once per step.

A step of length h with (8.9) can be seen as a concatenation of a step of length $h/2$ with the first order (1)[1] followed by a step of length $h/2$ with [1](1). Since both first-order methods are mutually adjoint, we conclude that, as pointed out in Section 3.6, (8.9) is a *symmetric* method.

Remark 8.3 The direct computation by the recipe in (3.34) of the tableaux of the concatenation of (1)[1] followed by [1](1) leads to

$$\begin{array}{|cc} 0 & 0 \\ 1/2 & 1/2 \\ \hline 1/2 & 1/2 \end{array}, \quad \begin{array}{|cc} 1/2 & 0 \\ 1/2 & 0 \\ \hline 1/2 & 1/2 \end{array}. \qquad (8.10)$$

This can be brought into the format (8.5) as follows. First we note that both q stages, $\mathbf{Q}_1$ and $\mathbf{Q}_2$ are identical, as the right tableau has two equal rows. Hence the expression defining $\mathbf{P}_2 = \mathbf{p}^{n+1}$ can be transformed from

$$\mathbf{p}^n + (h_{n+1}/2)[\mathbf{f}(\mathbf{Q}_1, t_n + h_{n+1}/2) + \mathbf{f}(\mathbf{Q}_2, t_n + h_{n+1}/2)]$$

into

$$\mathbf{p}^n + h_{n+1}\mathbf{f}(\mathbf{Q}_1, t_n + h_{n+1}/2).$$

After this the left tableau in (8.10) becomes

$$\begin{array}{|cc} 0 & 0 \\ 1 & 0 \\ \hline 1 & 0 \end{array}.$$

Now the stage $\mathbf{Q}_2$ plays no role, in view of the column of zeros in the last tableau. This makes it possible to replace the second row in the right tableau of (8.10) by any row vector without making any real difference to the method. The choice $(1/2, 1/2)$ as second row reveals that the composition (8.10) is in fact (8.9), as claimed above.

In the method (8.9) there are really two p stages and one q stage (cf. Remark 3.2). It would be possible to write it with a 2×1 a_{ij} matrix $(0,1)^T$ and a 1×2 A_{ij} matrix $(1/2, 0)$. However, and as discussed in Chapter 3, we throughout adhere to the convention that the numbers of stages for the p and q variables equal each other. □

Remark 8.4 In (8.9), the stage $\mathbf{Q}_1$ at the step $n \to n+1$ is an approximation to $\mathbf{q}(t_n + h_{n+1}/2)$. Upon denoting this intermediate

approximation by $q^{n+1/2}$, the formula defining p^{n+1} becomes
$$p^{n+1} - p^n = h_{n+1}f(q^{n+1/2}, t_n + h_{n+1}/2).$$
Furthermore, by subtracting the equation defining $q^{n-1/2}$ from the equation defining $q^{n+1/2}$ we find
$$q^{n+1/2} - q^{n-1/2} = \frac{h_n + h_{n+1}}{2}g(p^n).$$

Therefore (8.9) is equivalent to a *leap-frog* method, where the p variables are approximated at the grid points t_n and the q variables are approximated at the midway points $(t_n+t_{n+1})/2 = t_n+h_{n+1}/2$ (staggered grids). Observe that in the leap-frog version, there is no need to compute the q^n's. Also note that the leap-frog implementation makes it apparent that each step of the method only demands one evaluation of f and one evaluation of g. □

Along with (8.9) we consider
$$[1/2, 1/2](1, 0). \tag{8.11}$$

All the considerations made above in connection with (8.9) can obviously be translated to (8.11) by simply changing the roles of the p and q variables. In particular the method is the concatenation of [1](1) followed by (1)[1].

8.4.4 Specific methods: order 3

For order 3 there are five order conditions to be satisfied and $s = 3$ should leave a free parameter. The corresponding family (in the format (8.7)) was investigated by Ruth in his pioneering paper (1983). If we let B_1 be the free parameter, then it is found (Abia and Sanz-Serna (1993)) that B_2 has to be a root of the quadratic equation
$$12B_1B_2(B_1 + B_2) - 9(B_1^2 + 3B_1B_2 + B_2^2) + 12(B_1 + B_2) - 4 = 0,$$
and that
$$b_2 = \frac{3B_1 + 3B_2 - 2}{6B_1B_2}, \quad b_3 = \frac{-3B_1 + 2}{6B_2(B_1 + B_2)},$$
with b_1 and B_3 determined by the consistency requirements
$$\sum_i b_i = 1, \quad \sum_i B_i = 1.$$

Ruth suggested the choice $B_1 = 2/3$, which leads to a method with rational coefficients

$$[7/24, 3/4, -1/24](2/3, -2/3, 1). \tag{8.12}$$

Sanz-Serna (1992a) has considered an alternative member of this family, constructed as follows. A separable Hamiltonian system is left invariant by simultaneously interchanging the roles of kinetic and potential energies, coordinates and momenta and reversing the direction of t. It is easy to see that the same will be true for the numerical method

$$[b_1, b_2, b_3](B_1, B_2, B_3)$$

if

$$b_1 = B_3, \quad b_2 = B_2, \quad b_3 = B_1.$$

When these 3 equations are added to the 5 order conditions for order 3, a system of 8 equations with 6 unknowns results. This system turns out to have a unique real solution with

$$B_1 = 0.91966152 \tag{8.13}$$

a root of the quartic

$$12z^4 - 24z^2 + 16z - 3 = 0.$$

The remaining coefficients are determined by the formulae at the beginning of the present subsection.

Of course, along with the family of methods of type []() we have been discussing, it would be possible to consider methods of the format ()[]. For instance, from Ruth's method (8.12) we have both the procedure

$$(2/3, -2/3, 1)[7/24, 3/4, -1/24],$$

obtained by swapping the roles of the p and q variables, and the adjoint

$$(-1/24, 3/4, 7/24)[1, -2/3, 2/3]. \tag{8.14}$$

8.4.5 Specific methods: order 4 out of order 3

Sanz-Serna (1992a) suggested a means for having fourth-order integrations based on the order-3 methods discussed in the preceding subsection. It is enough to concatenate a step of length $h_{n+1}/2$ with a chosen method followed by a step of length $h_{n+1}/2$ with the adjoint method. The composition is a symmetric method, as we saw in Section 3.6, and hence has even order, i.e. order 4. (Actually

EXPLICIT PARTITIONED RUNGE-KUTTA METHODS 109

the results after the first substep of each step are order-4 approximations to the solution at $t_n + h_{n+1}/2$: up to t_n the integration has a global error $O(h^4)$, stepping from t_n to $t_n + h_{n+1}/2$ introduces the *local error* of the order-3 method, an additional $O(h^4)$ contribution.)

If the basic order-3 method is $(b_1, b_2, b_3)(B_1, B_2, B_3)$, then the concatenation with the adjoint is found to be, after manipulating the tableaux as in Remark 8.3,

$$(b_1, b_2, 2b_3, b_2, b_1, 0)[B_1, B_2, B_3, B_3, B_2, B_1].$$

The last q stage $\mathbf{Q}_6$ is never needed, due to the 0 weight on the left. Hence only five evaluations of $\mathbf{f}$ per step are needed. Furthermore, the left tableau presents an FSAL property (see Remark 5.1 and comments after (8.9)) and essentially only five evaluations of $\mathbf{g}$ per step are required.

Example 8.1 If the basic method is Ruth's adjoint scheme (8.14), then we have

$$(-1/24, 3/4, 7/12, 3/4, -1/24, 0)[1, -2/3, 2/3, 2/3, -2/3, 1],$$

a method we shall test in the next chapter. □

8.4.6 Specific methods: order 4

Of course, concatenating an order-3 method with its adjoint is not the only possible way in which order-4 methods can be constructed. For order 4, eight order conditions are necessary, see Table 7.2, and $s = 4$ should be sufficient in (8.5). The corresponding method was found by Forest and Ruth (1990) and Candy and Rozmus (1991):

$$(\omega, \nu, \omega, 0)[\omega/2, (\omega + \nu)/2, (\omega + \nu)/2, \omega/2], \qquad (8.15)$$
$$\omega = (2 + 2^{1/3} + 2^{-1/3})/3, \quad \nu = 1 - 2\omega.$$

As in other methods we have discussed before, the 0 weight implies that only three evaluations of $\mathbf{f}$ and $\mathbf{g}$ are required per step.

It is remarkable that the values of ω and ν here are those of b_1 and b_2 in (8.4). This parallelism is made even stronger if we notice that just as the order-4, diagonally implicit, symplectic method is a concatenation of implicit midpoint rule substeps of lengths $b_1 h_{n+1}$, $b_2 h_{n+1}$ and $b_1 h_{n+1}$, the method (8.15) is a concatenation of substeps of lengths ωh_{n+1}, νh_{n+1} and ωh_{n+1} with the leap-frog method (8.9). The reason for this parallelism is made apparent by the Lie theory presented in Chapters 12 and 13.

8.4.7 Specific methods: concatenations of the leap-frog methods

Later in the book we shall be considering methods that, like (8.15), are concatenations of leap-frog steps. It is appropriate to note at this stage that the concatenation of a step of length $\omega_1 h_{n+1}$, followed by a step of length $\omega_2 h_{n+1}$, ..., followed by a step of length $\omega_s h_{n+1}$ of the leap-frog method (8.9) is the $(s+1)$-stage method

$$(\omega_1, \omega_2, \ldots, \omega_s, 0) \left[\frac{\omega_1}{2}, \frac{\omega_1 + \omega_2}{2}, \ldots, \frac{\omega_{s-1} + \omega_s}{2}, \frac{\omega_s}{2}\right]. \quad (8.16)$$

A similar concatenation with the alternative leap-frog method in (8.11) leads to

$$\left[\frac{\omega_1}{2}, \frac{\omega_1 + \omega_2}{2}, \ldots, \frac{\omega_{s-1} + \omega_s}{2}, \frac{\omega_s}{2}\right] (\omega_1, \omega_2, \ldots, \omega_s, 0). \quad (8.17)$$

8.5 Available symplectic Runge-Kutta-Nyström methods

8.5.1 Implicit methods

Each symplectic RK scheme, when applied to a separable Hamiltonian with quadratic kinetic energy (6.12), leads, after elimination of the intermediate p stages, to an implicit, symplectic RKN method; cf. (3.33). We saw in Chapter 5 that this implied RKN method is more efficient than the original RK method.

8.5.2 Explicit methods: general format

There are explicit RKN methods that are symplectic. In fact, by combining the requirements of explicitness and symplecticness, the following family, first considered by Suris (1988), (1989), is easily found:

$$\begin{array}{c|cccc}
\gamma_1 & 0 & 0 & \cdots & 0 \\
\gamma_2 & b_1(\gamma_2 - \gamma_1) & 0 & \cdots & 0 \\
\vdots & \vdots & \vdots & \ddots & \vdots \\
\gamma_s & b_1(\gamma_s - \gamma_1) & b_2(\gamma_s - \gamma_2) & \cdots & 0 \\
\hline
 & b_1(1 - \gamma_1) & b_2(1 - \gamma_2) & \cdots & b_s(1 - \gamma_s) \\
\hline
 & b_1 & b_2 & \cdots & b_s
\end{array}. \quad (8.18)$$

With s stages there are $2s$ free parameters. Furthermore Okunbor and Skeel (1992b) prove that all explicit, symplectic RKN methods can be recast in the format (8.18), perhaps after reordering the stages and suppressing unnecessary stages.

Okunbor and Skeel (1992b) have pointed out that, for *implementation* purposes, (8.18) can be rewritten as an explicit PRK method, and hence only requires the storage of two d-dimensional vectors. More specifically, the s-stage (8.18) is the RKN method induced, as in (3.33), by the explicit, symplectic PRK method with $s+1$ stages

$$(b_1, b_2, \ldots, b_s, 0)[\gamma_1 - \gamma_0, \gamma_2 - \gamma_1, \ldots, \gamma_{s+1} - \gamma_s], \qquad (8.19)$$

where $\gamma_0 = 0$, $\gamma_{s+1} = 1$. Thus to implement (8.18) for (6.12), it is enough to implement the PRK method (8.19) as in the preceding section, taking of course into account that $\mathbf{g}(\mathbf{P}_i) = M^{-1}\mathbf{P}_i$, see (6.13).

Remark 8.5 Each explicit, symplectic RKN method arises from (and is best implemented as) an explicit, symplectic PRK method; why do we then consider explicit, symplectic RKN methods? When considering (8.18) (as the RKN it is) we have in mind the application to Hamiltonian problems with Hamiltonian functions of the special form (6.12). However, when looking at (8.19) as a PRK method, we are thinking of the application to *arbitrary* separable Hamiltonians (6.8). In general, the (RKN-)order of (8.18) is higher than the (PRK-)order of (8.19), as discussed in Remark 4.4.

This difference in order is only present for methods of order 4 or higher. For consistency both (8.18) and (8.19) require one condition, namely $\sum_i b_i = 1$. (Note that there are two consistency conditions for PRK methods, namely $\sum_i b_i = 1$, $\sum_i B_i = 1$. The second is automatically satisfied by (8.19). In a like manner, see Figs. 4.3 and 4.4 and Table 7.2, both (8.18) and (8.19) require an additional condition for order≥ 2, and two more for order ≥ 3. However at order 4, there are three order conditions for symplectic PRK methods and only two for symplectic RKN methods. □

8.5.3 Specific explicit methods

Each explicit, symplectic PRK method considered in the preceding section induces an explicit, symplectic RKN method. We specifically mention the RKN method induced by the leap-frog (8.9). This is the well-known Störmer-Verlet method

$$\begin{array}{c|c} 1/2 & 0 \\ \hline & 1/2 \\ \hline & 1 \end{array} \qquad (8.20)$$

used in many applications including as molecular dynamics (see e.g. Biesiadecki and Skeel (1992)).

In her thesis, M.P. Calvo (1992) (see also Calvo and Sanz-Serna (1992b), (1993a), (1993b)) developed an order-4, explicit, symplectic RKN method that it is not constructed through a PRK method and possesses some optimality properties. Calvo begins by noting that (8.18) possesses the FSAL property (see Remark 5.1) if $\gamma_1 = 0$, $\gamma_s = 1$. Hence with s stages there are $2s - 2$ free parameters in the corresponding tableau. For order 4, we need to impose (Table 7.2) six order conditions, so that four or more stages are required. Calvo chooses $s = 5$ (i.e., FSAL methods with four evaluations per step) in order to have two free parameters left after imposing the order conditions.

The two free parameters are chosen so as to minimize the coefficients of the leading term of the expansion of the local error. Recall from Section 4.6 that for an order-4 RKN method, and as far as the variables $\mathbf{v} = \dot{\mathbf{q}}$ are concerned, the $O(h^5)$ leading term in the local error is a sum, extended to the special Nyström rooted trees with five vertices (there are nine of them), of terms of the form

$$\frac{h^5}{5!}\alpha(\sigma\nu\rho\tau)\left(1 - \gamma(\sigma\nu\rho\tau)\Phi(\sigma\nu\rho\tau)\right) f(\sigma\nu\rho\tau)(\mathbf{v},\mathbf{q})$$

(in a Hamiltonian problem $\mathbf{v} = M^{-1}\mathbf{p}$). The elementary differential f depends on the problem being solved but not on the specific RKN method and it makes sense to choose, following Dormand and Prince, the method coefficients so as to minimize the nine quantities

$$\frac{1}{5!}\alpha(\sigma\nu\rho\tau)\left(1 - \gamma(\sigma\nu\rho\tau)\Phi(\sigma\nu\rho\tau)\right). \qquad (8.21)$$

(see e.g. Dormand and Prince (1989) and references therein). More precisely, if we denote by $\dot{A}$ the Euclidean norm in $\mathcal{R}^9$ of the vector with components (8.21), one tries to get methods with low values of $\dot{A}$.

Of course, the local error for the q variables should also be taken into account. For order-4 methods, this error is expressed in terms

AVAILABLE SYMPLECTIC RUNGE-KUTTA-NYSTRÖM METHODS 113

of the four special Nyström rooted trees with four vertices and gives rise to a 4-vector of error coefficients

$$\frac{1}{5!}\alpha(\sigma\nu\rho\tau)\left(1 - 5\gamma(\sigma\nu\rho\tau)\tilde{\Phi}(\sigma\nu\rho\tau)\right). \quad (8.22)$$

The Euclidean norm of this 4-vector is denoted by A. Calvo chooses the free parameters so as to minimize $\sqrt{A^2 + \dot{A}^2}$ and obtains the method

$$\begin{aligned}
\gamma_1 &= 0, \\
\gamma_2 &= 0.2051776615422863869, \\
\gamma_3 &= 0.6081989431465009739, \\
\gamma_4 &= 0.4872780668075869657, \\
\gamma_5 &= 1, \\
b_1 &= 0.0617588581356263250, \\
b_2 &= 0.3389780265536433551, \\
b_3 &= 0.6147913071755775662, \\
b_4 &= -0.1405480146593733802, \\
b_5 &= 0.1250198227945261338.
\end{aligned}$$

This and other methods are tested in the next chapter. Okunbor and Skeel (1993) give methods of orders 5 and 6.

CHAPTER 9

Numerical experiments

9.1 A comparison of symplectic integrators

9.1.1 Methods being compared

It is now time for comparing the various symplectic integrators we have discussed in the preceding chapter. We have implemented the following order-4 RKN methods (see Table 9.1):

Calvo The explicit, symplectic RKN method with five stages, four evaluations per step developed by Calvo, see Subsection 8.5.3. The formula has optimized error constants and has been implemented with constant step sizes.

Candy This is the explicit, symplectic RKN method induced (see (3.33)) by the four stages, three evaluations per step, PRK method (8.15). The implementation uses constant step sizes.

Dormand The explicit, *nonsymplectic* embedded pair described in Example 5.1. The algorithm requires three function evaluations per step and has been implemented (of course with variable step sizes) as described in Chapter 5. This algorithm is included in the comparisons as a good example of a state-of-the-art, optimized, standard integrator.

Gauss The *implicit,* symplectic RKN method induced by the two-stage Gauss method (see Example 3.3). This is implemented with constant step sizes, with the algebraic equations solved by functional iteration as described in Chapter 5. The algorithm requires two function evaluations per inner iteration; the number of evaluations per step depends then on the number of inner iterations required to solve the algebraic equations.

Ruth The explicit, symplectic RKN method induced by the PRK method with six stages and five function evaluations in Example 8.1. The implementation uses constant step sizes.

Table 9.1. *Fourth-order methods being compared*

Method	N	A	$\dot{A}$	$A \cdot N^5$	$\dot{A} \cdot N^5$
Calvo	4	7.10×10^{-4}	6.76×10^{-4}	0.73	0.69
Candy	3	3.69×10^{-2}	1.21×10^{-1}	8.99	29.41
Dormand	3	4.63×10^{-4}	1.75×10^{-3}	0.11	0.43
Gauss	–	3.24×10^{-3}	2.87×10^{-3}	–	–
Ruth	5	3.07×10^{-3}	3.43×10^{-3}	9.62	10.73
Sanz-Serna	5	1.88×10^{-3}	1.58×10^{-3}	5.88	4.94

Sanz-Serna This is similar to the algorithm Ruth. Rather than composing the adjoint (8.14) of Ruth's order-3 method (8.12) with Ruth's method, we here compose the adjoint of Sanz-Serna's method (8.13) with Sanz-Serna's method.

We use people's names to identify the algorithms in order to avoid labelling them as A1, A2, etc. or in another unhelpful way. It is not necessary to stress that, in the comparisons that follow, we only assess computer codes written by us, not the authors whose names are used as labels.

Table 9.1 displays, for each method, the number N of evaluations per step (this is the number of stages s, except for FSAL methods where it is $s - 1$) and the error constants A and $\dot{A}$ introduced in Subsection 8.5.3. Also shown are the *scaled error constants* AN^5 and $\dot{A}N^5$. Different methods require different numbers N of function evaluations per step and, to equalize work in a given experiment, one may use the methods with step sizes h proportional to N. Then the leading $O(h^5)$ term in the local error has constants proportional to AN^5 and $\dot{A}N^5$.

In the table we see that the methods Calvo and Dormand, derived to minimize the constants A and $\dot{A}$, have error constants of roughly the same size. However the symplectic method Calvo requires four evaluations per step, while the nonsymplectic method Dormand needs only three. This higher work is a consequence of symplecticness: to make the formula symplectic free parameters in the RKN tableau are sacrificed that otherwise could be used to decrease the size of the error constants. For this reason Dormand has better *scaled* error constants than Calvo.

The methods Gauss, Ruth and Sanz-Serna have error constants worse than Calvo and Dormand, but better than Candy. In fact, when scaled values are compared, it is clear that the methods la-

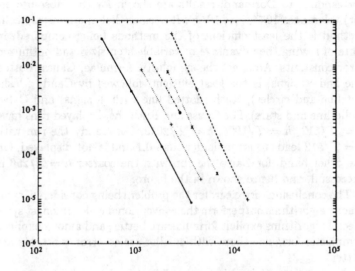

Figure 9.1. *Error as a function of CPU time after 21870 periods*

belled Ruth and Sanz-Serna look better than the method labelled Candy, in spite of the lower cost per step of the latter. Methods with fewer evaluations per step are then not necessarily better!

9.1.2 Results: Kepler's problem

The methods presented above were tested in Kepler's problem with initial condition (1.15). The eccentricity e was chosen to be 0.5. We recall from Section 5.3 that for this value of the eccentricity the method Dormand is clearly more efficient than it would be if implemented with constant step sizes. Thus the value of e tends to favour the nonsymplectic algorithm used in the comparison.

The methods were run up to final times $10\mathcal{T}$, $30\mathcal{T}$, $90\mathcal{T}$, ..., $21870\mathcal{T}$, where $\mathcal{T} = 2\pi$ is the period, with the error measured at the final point of the integration in the Euclidean norm of the $\mathcal{R}^4$ space of the variables (p_1, p_2, q_1, q_2). Runs with errors above 0.1 were disregarded and results for the method Sanz-Serna are not given as they are not significantly different from those of the related method Ruth.

Fig. 9.1 shows error against CPU time in seconds on a SUN Sparc IPX workstation. The rightmost, dashed line (plus signs) corresponds to Dormand; results are shown for the (absolute error) tolerances $10^{-9}, \ldots, 10^{-12}$. It is clear that the *nonsymplectic* method is the least efficient of the methods being compared, in spite of having the advantage of variable step sizes and optimized error constants. Amongst the symplectic formulae, Gauss (dotted line and × signs) is the least efficient, followed by Candy (dash-dot line and circles), Ruth (dotted line with + signs) and Calvo (solid line and stars). For Gauss and Ruth the displayed runs have $h = T/512$, $h = T/1024$, $h = T/2048$. For Candy the run with $h = T/512$ leads to errors larger than 0.1 and is not displayed. On the other hand, for Calvo the run with the coarser $h = T/256$ is successful, and hence shown in the figure.

The conclusions are clear: for the problem being considered, symplectic algorithms outperform the conventional code; amongst symplectic algorithms explicit formulae are better, and among explicit, symplectic formulae those with smaller scaled error constants are better.

For reasons explained in Section 5.3, it is also useful to measure cost in terms of function evaluations. This is done in Fig. 9.2 for the runs in Fig. 9.1. The only method whose relative position changes is Candy; this is now the least efficient, while before it was third. With work measured by number of function evaluations the overheads in the variable step Dormand and in the implicit Gauss are not measured and Dormand and Gauss overtake Candy. In particular, Candy would be less efficient than Dormand in problems with expensive evaluations: symplecticness cannot make up for the large error constants of this method. The symplectic methods Gauss and Ruth do slightly better than the automatic code Dormand. This is remarkable because Gauss has the drawback of being implicit and both Gauss and Ruth use constant step sizes. The method labelled Calvo is clearly the best. Its efficiency roughly doubles that of the code Dormand. This shows the benefits of symplectic methods with free parameters and error-constant optimization when compared with other symplectic methods.

The poor relative performance of the nonsymplectic method, in spite of its small error constants, suggests that there is something in the *error propagation* mechanism of symplectic methods that gives them an advantage that may make up for their larger error constants (i.e., for their larger local errors). In connection with this, it is useful to consider Fig. 9.3, which gives error as a function of time in a log-log scale. The runs shown correspond (from bottom to top) to Dormand with a tolerance 10^{-9} (dashed line

A COMPARISON OF SYMPLECTIC INTEGRATORS

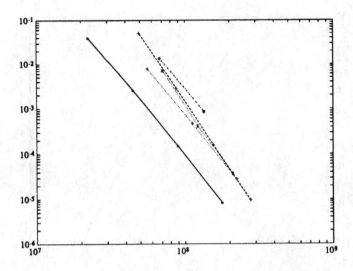

Figure 9.2. *Error as a function of number evaluations after 21870 periods*

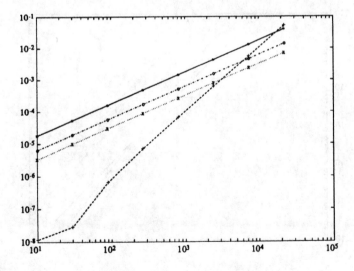

Figure 9.3. *Error as a function of t*

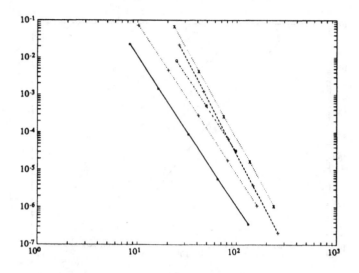

Figure 9.4. *Error as a function of CPU time after 810 periods*

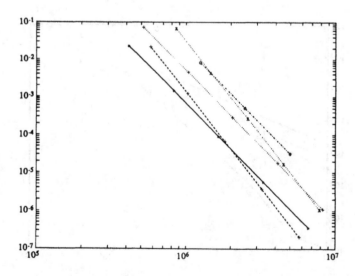

Figure 9.5. *Error as a function of number of evaluations after 810 periods*

and + signs), Gauss with $h = T/512$ (dotted line and × signs), Candy with $h = T/1024$ (dash-dot line and circles) and Calvo $h = T/256$. No line for Ruth is shown in order to get a cleaner figure. Since different methods are working differently, no information as to the relative efficiency can be derived from this figure (relative efficiency has already been discussed). It is the slope of each line that is relevant. The point we want to make is that while the error grows like t (slope 1) for symplectic methods, it grows like t^2 (slope 2) for the standard, nonsymplectic code: symplecticness pays in long integrations. We have analysed the error propagation mechanism in Kepler's problem in Calvo and Sanz-Serna (1992b), (1993a) (1993b) and further details cannot be given here. Of course we make no claim that symplectic methods would have linear error growth in *all* Hamiltonian problems.

Fig. 9.4 differs from Fig. 9.1 in that now the final integration time is $810T$ rather than $21870T$. A comparison of both figures reveals that, within the class of symplectic algorithms, the relative efficiency does not change with the final integration time. However the nonsymplectic Dormand comes out better at $810T$ than it does at $21870T$: it is now slightly better than Gauss and comparable to Candy. In Fig. 9.4 the symplectic methods are run with $h = T/128$, $h = T/256$, ... (except for Candy that requires $h = T/512$ or smaller to produce errors below 0.1). The variable-step code is run with tolerances of $10^{-7}, 10^{-8}, \ldots$

Fig. 9.5 also corresponds to a final time of 810 periods, but now efficiency is measured in number of function evaluations. A comparison with Fig. 9.2 is in order. At time $t = 810T$ the standard code is as efficient, in terms of function evaluations, as the best symplectic method being compared. The main interest of symplectic integration is in long integrations; in short integrations small local errors are of paramount importance and the parameters in the method should be directed at decreasing the error constants rather than at rendering the formula symplectic.

9.1.3 Results: Hénon-Heiles problem

Our second test problem is qualitative. In Chapter 1 we saw that the solution of the Hénon-Heiles problem represented in Fig. 1.6 is quasiperiodic; in the Poincaré section the intersections lie on a curve. We now run for $0 \leq t \leq 200000$ the methods being considered and see how much work each method requires to correctly identify that the intersections remain on a curve. To this end, we

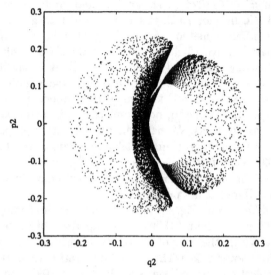

Figure 9.6. *A Poincaré section computed by the method Dormand with tolerance* 10^{-4}

run the constant-step algorithms with $h = 1, 1/2, 1/4, \ldots$ and stop halving h when the correct section is produced. For the variable-step algorithm Dormand we use tolerances $10^{-4}, 10^{-5}, \ldots$

Fig. 9.6 gives the section as computed by the algorithm Dormand with tolerance 10^{-4}. Only every third intersection is plotted to get a cleaner picture. Clearly the algorithm does not bear out the quasiperiodicity of the solution. It is necessary to reduce the tolerance to 10^{-6} to come up with the right qualitative behaviour. For this value of the tolerance, the algorithm requires over 5,000,000 function evaluations and a CPU time of roughly 250 seconds on a SUN Sparc IPX workstation (see Table 9.2).

In this problem the algorithm Calvo is an order of magnitude better than the nonsymplectic code. In fact, Calvo identifies the right behaviour with $h = 1$, requiring only 800,000 function evaluations and 19 seconds of CPU time.

In Table 9.2 we also see that Ruth is slightly worse than Calvo. The algorithm Sanz-Serna performs, as before, like Ruth. Candy is again worse than Ruth or Sanz-Serna: the advantage of fewer evaluations per step is offset by the smaller value of the step size required, $h = 1/2$. The performance of Gauss is disappointing: the

Table 9.2. *Cost of the cheapest successful run*

Method	$h/$Tol	Evaluations	CPU
Calvo	1	800001	19
Candy	1/2	1200000	27
Dormand	10^{-6}	5088778	245
Gauss	1/4	9621145	247
Ruth	1	1000000	22
Sanz-Serna	1	1000000	22

iteration employed to solve the algebraic equations fails for $h = 1$ and $h = 1/2$. For $h = 1/4$ the algorithm succeeds, but then is as expensive as the nonsymplectic Dormand.

Remark 9.1 In the results reported the CPU time devoted to the computation of intersections for the Poincaré section is not included. This is method-independent and, roughly, equals 70 seconds. To find the Poincaré section we check whether at the time step from t_n to t_{n+1} the variable q_1 changes sign. If it does, we replace the function $q_1(t)$ in the interval $[t_n, t_{n+1}]$ by a Hermite cubic interpolant, i.e., we find a cubic polynomial that matches q_1 and $\dot q_1 = p_1$ at t_n and t_{n+1}. We then find the root $t_{section}$ of this polynomial by Newton's method. The point (p_2, q_2) to be plotted is obtained by evaluating at $t_{section}$ the Hermite cubic interpolants of the functions p_2 and q_2. □

9.1.4 Results: computation of frequencies

In the last experiment in this section the aim is to find the frequencies of a quasiperiodic solution. We consider the modified Kepler problem of Chapter 1, with initial condition as in Fig. 1.3. We saw that to generate this picture it is necessary to start from a time-series $q_1(0), q_1(1), \ldots, q_1(2^{13} - 1)$. These values are numerically generated by integrating the equations of motion for $0 \leq t \leq 2^{13} - 1 = 8191$. For simplicity only two methods are considered. The first is the algorithm Calvo, the symplectic algorithm that performed best in the previous experiment. This is run with $h = 1/4, 1/8, \ldots$ The second is the nonsymplectic, order-4 method of the pair in Example 5.1 used with constant steps $h = 1/4, 1/8, \ldots$ We employ this rather than the variable-step embedded pair Dormand, because the need to output the solution

Table 9.3. *Frequencies in the quasiperiodic solution*

h	Second line		Third line	
	Symp.	Nonsymp.	Symp.	Nonsymp.
1/4	0.1835	***	0.3621	***
1/8	0.1837	***	0.3623	***
1/16	0.1837	0.1841	0.3623	0.3629
1/32	0.1837	0.1837	0.3623	0.3624
1/64	0.1837	0.1837	0.3623	0.3623

values at $t = 1, 2, \ldots, 8191$ interferes with the step size selection mechanism or demands interpolation (Hairer *et al.* (1987), Section II.5).

Recall from Fig. 1.3 that there are spectral lines at three frequencies $\nu = 0.0051, 0.1837, 0.3623$. For reasons of simplicity we only report on the results corresponding to the two largest frequencies. The results are contained in Table 9.3. For $h = 1/4$ or $h = 1/8$ the nonsymplectic algorithm does not bear out the quasiperiodicity of the solution. The corresponding periodograms have continuous spectra of frequencies, rather than well-defined lines. On the other hand, the symplectic algorithm identifies the quasiperiodicity of the solution even for the coarse value $h = 1/4$. For this value of h the errors in the frequencies for the symplectic algorithm are very small, of about 0.1%. For smaller values of h the frequencies are correct. In this example the symplectic algorithm can use step sizes at least four times larger than those needed by the nonsymplectic method.

9.2 Variable step sizes for symplectic methods

Most numerical results for symplectic integrators reported in the literature refer to constant step size implementations. Standard nonsymplectic methods were also developed first in constant h implementations; more sophisticated (nonsymplectic) software appeared later.

The present authors (Calvo and Sanz-Serna (1993b), see also Calvo (1992), Calvo and Sanz-Serna (1992b), Calvo and Sanz-Serna (1993a)) attempted the construction of a variable step size code, based on an embedded RKN pair of orders 3 and 4. The order-4 method is the formula presented in Subsection 8.5.3 and used

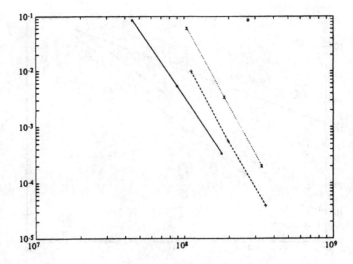

Figure 9.7. *Error as a function of number of evaluations after 21870 periods*

in the experiments above. The embedded order-3 method is constructed by optimization techniques due to Dormand and Prince; the formula coefficients can be seen in the papers by Calvo and Sanz-Serna referred to above.

The practical experience with the variable-step symplectic code was most disappointing. The code did not show any of the advantages of the symplectic constant step size algorithms tested in the preceding section. It performs just as a standard variable-step code, with the drawback that it has unfavourable error constants, because in its construction degrees of freedom are used to ensure symplecticness that could otherwise have been employed to reduce the error constants (compare the entries Calvo and Dormand in Table 9.1).

Fig. 9.7 is an efficiency plot for the Kepler test problem we have been employing in this chapter (work is measured in number of function evaluations; using CPU time does not change the conclusions to follow). The most efficient method is the constant step size Calvo used in the experiments of the preceding section (stars joined by a solid line; shown are runs for $h = 2\pi/512$, $2\pi/1024$, $2\pi/2048$). This method is followed by the variable h Dormand

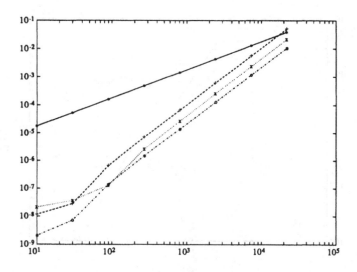

Figure 9.8. *Error as a function of t*

(crosses joined by a dashed line; values of the tolerance 10^{-10}, 10^{-11}, 10^{-12}, 10^{-13}). The × signs joined by a dotted line correspond to the variable-step symplectic method. We see clearly that adding step-change facilities to a symplectic formula results in a *degraded performance*, just the opposite to the situation with standard formulae (see Fig. 5.5). The relative performance of the variable-step implementations of the formulae by Dormand et al. and Calvo can be accounted for in terms of local accuracy: in Table 9.1 we see that the error constants for both formulae are roughly of the same size but that the costs per step are in a ratio 3/4. This is precisely the efficiency ratio observed in the figure. Finally the circle in the figure corresponds to the Dormand formula implemented with constant step size $h = 2\pi/4096$. A comparison with the results for the variable h implementation of the same formula confirms once more that, for the nonsymplectic formulae, variable h is a great help.

Fig. 9.8 shows error growth as a function of t. The main point to observe is that, with variable step sizes, the symplectic formula exhibits a *quadratic* error growth, i.e., behaves in a nonsymplectic way.

Further discussion of these numerical experiments can be seen in the references by Calvo and Sanz-Serna cited above. For more literature discouraging the use of variable step sizes with symplectic formulae see Gladman *et al.* (1991), Skeel and Gear (1992), Gear (1992), Okunbor (1992), Skeel (1993). In the next chapter we will examine these issues from a more theoretical point of view.

9.3 Conclusions and recommendations

In this chapter we have tested several order-4 RKN algorithms. For short-time integrations conventional software is to be preferred; symplecticness plays no role and a good method is one with small local error constants. On the other hand, for long-time integration, the symplectic formulae used with *constant step sizes* outperform variable-step algorithms implementing good standard (nonsymplectic) formulae. This is true both in cases where accurate solutions are required (Kepler test problem) and in cases where the integration is only performed to ascertain some qualitative features of the solution (Hénon-Heiles test problem).

Of course, not all symplectic formulae are equally efficient. Formulae with small *scaled* error constants are to be preferred. Good formulae are not those that work least per step; having more stages than the minimum required for a given order is not bad if this leads to smaller scaled error constants.

Amongst the symplectic formulae tested, the optimized scheme due to Calvo is clearly the best. It is not unlikely that further optimization is possible and that even better formulae may be obtained by carefully tuning the tableau coefficients.

Symplectic formulae are to be used with constant step sizes. With variable step sizes they do not behave as symplectic integrators and one is then better off using a good standard method.

CHAPTER 10

Properties of symplectic integrators

10.1 Backward error interpretation

10.1.1 An example

In our opinion, the most appealing feature of symplectic integration is the possibility of *backward error interpretation*. Of course the principle of backward error analysis is well-established in numerical linear algebra and other fields. For numerical differential equations see Sanz-Serna (1992b), Eirola (1993), Hairer (1993), Sanz-Serna and Larsson (1993). The idea is very similar to the *method of modified equations*, see Warming and Hyett (1974) and, for a more rigorous treatment, Griffiths and Sanz-Serna (1986). For simplicity, we restrict our attention in this section to *autonomous* Hamiltonian problems (see Remark 4.1).

Let us begin with an example: the one-degree-of-freedom separable Hamiltonian $H = T(p) + V(q)$, leading to the system

$$dp/dt = f(q), \qquad dq/dt = g(p), \qquad (10.1)$$

where $f = -V'$ and $g = T'$. We assume that $f(0) = 0$ and $g(0) = 0$ so that the origin is an equilibrium of (10.1). Also assume that $T''(0) = g'(0) > 0$ and $V''(0) = -f'(0) > 0$, so that both T and V have minima at 0. Then the origin is a (stable) centre for (10.1).

We integrate (10.1) by the first-order symplectic PRK method (1)[1] of Chapter 8:

$$p^{n+1} = p^n + hf(q^{n+1}), \qquad q^{n+1} = q^n + hg(p^n). \qquad (10.2)$$

In order to describe the behaviour of the points (p^n, q^n) computed by (10.2), we could just say that they approximately behave like the solutions $(p(t_n), q(t_n))$ of (10.1). This would not be a very precise description because the true flow $\phi_{h,H}$ and its numerical approximation $\psi_{h,H}$ differ in $O(h^2)$ terms. Can we find *another*

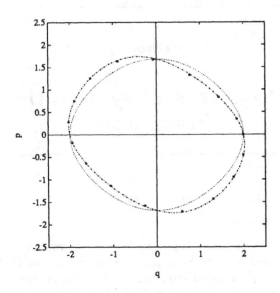

Figure 10.1. *Computed points, true trajectories and modified trajectories for the pendulum*

differential system S_2 so that (10.2) is consistent of the *second* order with S_2? The points (p^n, q^n) would then be closer to the solutions of S_2 than to the solutions of the system (10.1) we want to integrate. To find the *modified* system S_2 use an *ansatz*

$$dp/dt = f(q) + hF(p,q), \qquad (10.3)$$
$$dq/dt = g(p) + hG(p,q),$$

(note that h features here as a parameter so that $S_2 = S_2^h$), substitute the solutions of the modified system (10.3) into the difference equations (10.2) and ask for an $O(h^3)$ residual. This leads to the following expression for (10.3)

$$dp/dt = f(q) + \frac{h}{2}g(p)f'(q), \qquad (10.4)$$

$$dq/dt = g(p) - \frac{h}{2}g'(p)f(q),$$

a Hamiltonian system with Hamiltonian

$$H_2^h = T(p) + V(q) + (h/2)T'(p)V'(q).$$

Figure 10.1 corresponds to the pendulum equations (Chapter 1)

$g(p) = p$, $f(q) = -\sin q$ with initial condition $p(0) = 0$, $q(0) = 2$. The stars plot the numerical solution with $h = 0.5$. The dotted line $H = constant$ provides the true pendulum solution. The dash-dot line $H_2 = constant$ provides the solution of the modified system (10.4). The agreement of the computed points with the modified trajectory is very good.

If we are not satisfied with S_2^h, we can find a differential system S_3^h for which (10.2) is consistent of the third order. Again S_3^h turns out to be a Hamiltonian problem; the expression for the Hamiltonian is

$$H_3(h) = T(p) + V(q) + (h/2)T'(p)V'(q) \\ + (h^2/12)[T'''(p)V'(q)^2 + T'(p)^2V'''(q)]. \quad (10.5)$$

There is no limit: we shall see (Example 12.2) that, for any positive integer ρ, a Hamiltonian system $S_{H_\rho^h}$ can be found such that the method $\psi_{h,H}$ differs from the flow ϕ_{h,H_ρ^h} in terms of order $O(h^{\rho+1})$. By going from local to global errors, in any bounded time interval, the computed points are $O(h^\rho)$ away from the solution of $S_{H_\rho^h}$.

What is the situation when using a nonsymplectic method? Take the standard forward Euler method as an illustration. Again a modified system S_2^h can be found for which consistency is of the second order. This now reads

$$dp/dt = \left[f(q) + \frac{h}{2}g(p)f'(q)\right] - hg(p)f'(q),$$

$$dq/dt = \left[g(p) - \frac{h}{2}g'(p)f(q)\right];$$

the terms in brackets replicate the Hamiltonian system (10.4), but there is an extra term $-hg(p)f'(q)$. Since $g(0) = 0$, we have, to first order in p, that $-hg(p)f'(q) \simeq -hg'(0)pf'(q)$; but $f'(q)$ is negative for q small and hence $-hg(p)f'(q)$ is a force that has the same sign as the momentum p. We are therefore in the presence of *negative dissipation*. In any bounded interval the computed points are $O(h)$ away from the solution of the Hamiltonian system (10.1) we want to solve, but $O(h^2)$ away from the solutions of a system where the Hamiltonian character has been lost and the origin is an unstable spiral point.

10.1.2 The general case

Even though the considerations above have been presented by means of an example, they hold for all symplectic methods $\psi_{h,H}$: *for any autonomous Hamiltonian system with a smooth Hamiltonian H and any positive integer ρ, a modified autonomous Hamiltonian system $S_{H_\rho^h}$ can be found, such that $\psi_{h,H}$ differs from the flow ϕ_{h,H_ρ^h} in $O(h^{\rho+1})$ terms.* The difference between the true Hamiltonian H and the modified Hamiltonian H_ρ^h is $O(h^r)$, with r the order of the method. If the method $\psi_{h,H}$ belongs to any of the standard classes of methods (including RK, PRK and RKN methods) this result is proved by Hairer (1993). For general symplectic transformations ψ_h depending on a small parameter h with $\psi_{h=0} = I$ see, e.g., Dragt and Finn (1976), Theorem 2. (In this reference the successive terms in H_ρ^h do not correspond to powers $h^0, h^1, \ldots, h^{\rho-1}$ but rather to homogeneous terms of degree $2, 3, \ldots, \rho + 1$ in the variables p_i, q_i; the technique there can be however adapted to the case at hand.)

As illustrated in the example of the preceding subsection, the functions H_ρ^h, $\rho = 2, 3, \ldots$ are truncations of a series H_∞^h in powers of h. If this power series converges and furthermore $\psi_{h,H} = \phi_{h,H_\infty^h}$, then the computed points *exactly* lie on the solutions of the system S_∞^h with Hamiltonian H_∞^h.

Example 10.1 Consider $H = p^2/2 + q^2/2$ integrated by method (1)[1] as in the preceding subsection. The modified Hamiltonian system S_∞^h is given by (Beyn (1991), p. 221)

$$\frac{d}{dt}\begin{bmatrix} p \\ q \end{bmatrix} = \left(h^{-1} \log \begin{bmatrix} 1 - h^2 & -h \\ h & 1 \end{bmatrix}\right) \begin{bmatrix} p \\ q \end{bmatrix}. \tag{10.6}$$

To compute the time h flow of this system one exponentiates h times the matrix in round brackets. The equations (10.2) are then recovered. □

In general, for *nonlinear* problems, the series for H_∞^h does not converge and it is *impossible* to find an autonomous differential system so that the computed points exactly lie on trajectories of the system (Sanz-Serna (1991), p. 168). However, if H is sufficiently well behaved it can be shown (Neishtadt (1984), cf. Lasagni (1990), MacKay (1992)) that by retaining for each $h > 0$ a large but finite number of terms $N = N(h)$ of the series for H_∞^h, a Hamiltonian $\tilde{H}^h$ can be obtained for which the corresponding h-flow differs from $\psi_{h,H}$ in terms that tend to 0 exponentially fast as $h \to 0$.

In any case the conclusion is the same: *for a symplectic integrator applied to an autonomous Hamiltonian system, modified autonomous Hamiltonian problems exist so that the computed points lie either exactly or 'very approximately' on the exact trajectories of the modified problem.* This makes possible a backward error interpretation of the numerical results: the computed solutions are solving exactly (or 'very approximately') a nearby Hamiltonian problem. In a modelling situation where the exact form of the Hamiltonian H may be in doubt, or some coefficients in H may be the result of experimental measurements, the fact that integrating the model numerically introduces perturbations to H comparable to the uncertainty in H inherent in the model is the most one can hope for.

On the other hand, when a nonsymplectic formula is used the modified system is not Hamiltonian: the process of numerical integration perturbs the model in such a way as to take it out of the Hamiltonian class. Of course, the acceptability of such non-Hamiltonian perturbations is a question that should be decided in each individual modelling problem.

A further important point to be noted is as follows: *the backward error interpretation only holds if the numerical solution at time t_n is computed by iterating n times one and the same symplectic map.* If, alternatively, one composes n symplectic maps (one from t_0 to t_1, a different one from t_1 to t_2, etc.) the backward error interpretation is lost. This is illustrated in the next subsection.

10.1.3 Application to variable-step sizes

It is appropriate to relate the failure of symplectic variable-step algorithms to the backward error analysis above. Let $\psi_{h,H}$ be a symplectic integrator for the autonomous Hamiltonian system S_H. If ρ is a large positive integer, construct H_ρ^h such that $\psi_{h,H} = \phi_{h,H_\rho^h} + O(h^{\rho+1})$. If h is held constant during the integration, the initial condition is numerically advanced from $t_0 = 0$ to $t = t_n$ by the mapping

$$\overbrace{\psi_{h,H}\psi_{h,H}\cdots\psi_{h,H}}^{n},$$

which for t_n in a bounded time interval differs in $O(h^\rho)$ terms from the composition

$$\overbrace{\phi_{h,H_\rho^h}\phi_{h,H_\rho^h}\cdots\phi_{h,H_\rho^h}}^{n} = \phi_{t_n,H_\rho^h}$$

(see (2.3)); the computed points stay very close to the points $\phi_{t_n, H_\rho^h}(\mathbf{p}^0, \mathbf{q}^0)$ on the solution of $S_{H_\rho^h}$ with initial values $(\mathbf{p}^0, \mathbf{q}^0)$. The situation is quite different for variable step sizes. Now the initial condition is advanced by

$$\psi_{h_n, H} \psi_{h_{n-1}, H} \cdots \psi_{h_1, H}, \tag{10.7}$$

an approximation to

$$\phi_{h_n, H_\rho^{h_n}} \phi_{h_{n-1}, H_\rho^{h_{n-1}}} \cdots \phi_{h_1, H_\rho^{h_1}}.$$

The last expression cannot be interpreted as the t_n-flow of a Hamiltonian problem: the Hamiltonian functions being used at different steps are different. This shows that the backward error interpretation of symplectic integration does not hold for variable step sizes.

There is difficulty here: for a given value of the tolerance, in a variable step-size code the step points t_n are actually functions of the initial point $(\mathbf{p}^0, \mathbf{q}^0)$ (and also of the initial guess for the first step size). Therefore the algorithm does not really effect a transformation mapping the phase space Ω at $t = 0$ into the phase space Ω at time t. Rather $(\Omega \times (t = 0))$ is mapped into some curved $2d$-dimensional surface in the $(2d+1)$-dimensional extended phase space. It is then possible to question the relevance of (10.7) to the analysis of the variable step implementation. However in the experiments in Section 9.2 only one fixed initial condition was used so that, in a 'mental experiment', one could pretend that the sequence of step sizes $h_1, h_2, \ldots$ actually used in the integration was recorded and would have been used to integrate neighbouring initial conditions (see also Hairer et al. (1987), end of Section II.5). In this context, compatible with the numerical experiments, the initial condition is really advanced by the symplectic transformation (10.7).

10.2 An alternative approach

If

$$(\mathbf{p}, \mathbf{q}) = \psi_{h, H}(\mathbf{p}^0, \mathbf{q}^0) \tag{10.8}$$

is a numerical method for an *autonomous* Hamiltonian problem, it is a simple matter to find a *nonautonomous* differential system $\hat{S}$ satisfied by the functions $(\mathbf{p}(h), \mathbf{q}(h))$: we differentiate (10.8) with respect to h and eliminate $(\mathbf{p}^0, \mathbf{q}^0)$ in the result by using (10.8).

Example 10.2 For the symplectic method (10.2) this procedure yields the system $\hat{S}$

$$\frac{dp}{dh} = f(q) + hf'(q)\,g(p - hf(q)),$$
$$\frac{dq}{dh} = g(p - hf(q)),$$

which (this should not be surprising by now) is a Hamiltonian system, with Hamiltonian

$$\hat{H}(p,q,h) = T(p + hV'(q)) + V(q)$$
$$= T(p) + V(q) + hT'(p)V'(q) + (h^2/2)T''(p)V'(q)^2 + \ldots$$

Since h is our 'time' the system is nonautonomous. The mapping $\psi_{h,H}$ of the method (10.8) is the solution operator $0 \to h$ of the nonautonomous system with Hamiltonian $\hat{H}$. The fact that $\hat{H} = H + O(h)$ reflects the first-order accuracy of the method. □

Let us prove that if (10.8) is symplectic, the system $\hat{S}$ is in fact a Hamiltonian system. Differentiation in (2.7) with respect to 'time' h yields

$$\frac{\partial(\dot{y})^T}{\partial(y^0)} J \frac{\partial(y)}{\partial(y^0)} + \frac{\partial(y)^T}{\partial(y^0)} J \frac{\partial(\dot{y})}{\partial(y^0)} = 0,$$

an identity that we multiply on the left by $(\partial(y^0)/\partial(y))^T = (\partial(y)/\partial(y^0))^{-T}$ and on the right by $\partial(y^0)/\partial(y) = (\partial(y)/\partial(y^0))^{-1}$. This leads to

$$\frac{\partial(y^0)^T}{\partial(y)} \frac{\partial(\dot{y})^T}{\partial(y^0)} J + J \frac{\partial(\dot{y})}{\partial(y^0)} \frac{\partial(y^0)}{\partial(y)} = 0$$

or, after using the chain rule and the equality $J^T = -J$,

$$-\frac{\partial(J\dot{y})^T}{\partial(y)} = \frac{\partial(J\dot{y})}{\partial(y)}.$$

Thus the Jacobian matrix $\partial(J\dot{y})/\partial(y)$ is symmetric and (if the phase space Ω is simply connected) there exists a scalar function $\hat{H}$ such that $J\dot{y} = \nabla \hat{H}$, i.e. $\dot{y} = J^{-1}\nabla\hat{H}$, as we required.

Using the notation of Section 2.1

$$\psi_{h,H} = \Phi_{\hat{H}(\mathbf{p},\mathbf{q},t)}(t = h, t = 0).$$

What is unsatisfactory with this approach is that taking two steps $0 \to h \to 2h$ with the numerical method is *not* going from $t = 0$ to $t = 2h$ with $S_{\hat{H}}$:

$$\psi_{t,H}\psi_{t,H} = \Phi_{\hat{H}(\mathbf{p},\mathbf{q},t)}(t = h, t = 0)\Phi_{\hat{H}(\mathbf{p},\mathbf{q},t)}(t = h, t = 0)$$
$$\neq \Phi_{\hat{H}(\mathbf{p},\mathbf{q},t)}(t = 2h, t = 0)$$

(cf. 2.2). There is a way around this problem: for $0 \leq t < h$ we keep the Hamiltonian $\hat{H}(\mathbf{p}, \mathbf{q}; t)$ found above and for $h \leq t < 2h$, $2h \leq t < 3h, \ldots$ we repeat it periodically. The good news is that now the nonautonomous system is such that

$$\Phi_{\hat{H}}(t = nh, t = 0) = \Phi_{\hat{H}}(t = h, t = 0)^n$$

(this is a consequence of the semigroup property of the period map of a periodic differential system, Arnold (1989) 25B). Hence, the numerically computed points lie exactly on solutions of this nonautonomous system. The bad news is that the new $\hat{H}$ is not only nonautonomous, but also discontinuous as a function of t. Such a lack of smoothness is of course not very welcome.

The Hamiltonian $\hat{H}$ can be explicitly found by means of generating functions, see Sanz-Serna (1992c), Section 12.

McLachlan and Atela (1992) use discrepancy between H and $\hat{H}$ as $t \to 0$ to measure the accuracy of the method $\psi_{h,H}$. This essentially coincides with the canonical theory of the order to be studied in Chapter 11.

10.3 Conservation of energy and other invariant quantities

10.3.1 Exact conservation: positive results

We saw in Section 1.1 that for autonomous Hamiltonian problems H is a conserved quantity: $H(\mathbf{p}(t), \mathbf{q}(t))$ does not vary with t if $(\mathbf{p}(t), \mathbf{q}(t))$ is a solution of the system. Do symplectic integrators possess the analogous property that, except for rounding errors, $H(\mathbf{p}^n, \mathbf{q}^n)$ does not vary with n along a numerically computed solution? More generally, if I is a conserved quantity for an autonomous Hamiltonian system $(I((\mathbf{p}(t), \mathbf{q}(t)) = constant)$ is I also conserved by symplectic integrators?

For the case where I is *quadratic* (for instance I is the energy in the harmonic oscillator or the angular momentum (1.8) in Kepler's problem), we have the following result, essentially due to Cooper (1987) (see Sanz-Serna (1988)).

Theorem 10.1 *Assume that $I(\mathbf{y}) = (1/2)\mathbf{y}^T S \mathbf{y}$, with S a constant symmetric matrix, is a conserved quantity of a (not necessarily Hamiltonian) autonomous system (3.1). Then I is also conserved when the system is integrated by a symplectic RK method, i.e. $(1/2)\mathbf{y}^{nT} S \mathbf{y}^n$ does not vary with n.*

CONSERVATION OF ENERGY

Proof. From
$$0 = (d/dt)I(\mathbf{y}(t)) = \mathbf{y}(t)^T S \dot{\mathbf{y}}(t) = \mathbf{y}(t)^T S \mathbf{F}(\mathbf{y}(t))$$
we see that $\mathbf{y}^T S \mathbf{F}(\mathbf{y}) \equiv 0$.
According to (3.12)

$$I(\mathbf{y}^{n+1}) = I(\mathbf{y}^n) + h_{n+1} \sum_{i=1}^{s} b_i \mathbf{F}(\mathbf{Y}_i)^T S \mathbf{y}^n$$
$$+ \frac{h_{n+1}^2}{2} \sum_{i,j=1}^{s} b_i b_j \mathbf{F}(\mathbf{Y}_i)^T S \mathbf{F}(\mathbf{Y}_j).$$

We now use (3.14) to eliminate $\mathbf{y}^n$ from the first sum. This leads, after rearranging, to

$$I(\mathbf{y}^{n+1}) = I(\mathbf{y}^n) + h_{n+1} \sum_{i=1}^{s} b_i \mathbf{F}(\mathbf{Y}_i)^T S \mathbf{Y}_i$$
$$+ \frac{h_{n+1}^2}{2} \sum_{i,j=1}^{s} (b_i b_j - b_i a_{ij} - b_j a_{ji}) \mathbf{F}(\mathbf{Y}_i)^T S \mathbf{F}(\mathbf{Y}_j).$$

The first sum vanishes, because as noted above $\mathbf{F}(\mathbf{Y}_i)^T S \mathbf{Y}_i = 0$. The second sum is 0 by the symplecticness condition (6.5). □

In an analogous way one can prove:

Theorem 10.2 *Assume that $I(\mathbf{p}, \mathbf{q}) = \mathbf{p}^T S \mathbf{q}$ with S a constant matrix is a conserved quantity of a (not necessarily Hamiltonian) autonomous partitioned system (3.23). Then I is also conserved when the system is integrated by a symplectic PRK method, i.e., $I(\mathbf{p}^n, \mathbf{q}^n)$ does not vary with n.*

A similar result holds for symplectic RKN methods.

Theorem 10.1 implies in particular that if a *linear* autonomous system with (quadratic) Hamiltonian H is integrated by a symplectic RK method, then H is exactly conserved along numerical trajectories.

10.3.2 Exact conservation: negative results

If we still assume linearity in the system, but now we use a symplectic PRK or RKN method, conservation of H does not hold in general. This is because the format $I(\mathbf{p}, \mathbf{q}) = \mathbf{p}^T S \mathbf{q}$ covered by Theorem 10.2 is not of course the most general quadratic function of the variables $\mathbf{p}$ and $\mathbf{q}$.

Example 10.3 We saw in Example 10.1 that, for the harmonic oscillator $H = (1/2)(p^2 + q^2)$ integrated by the method (1)[1], the computed points lie exactly on trajectories of the modified system (10.6) and hence on the lines $H_\infty^h(p, q) = constant$ in the (p, q) plane. For h small, these lines are ellipses, while for conservation of energy we wanted the points to be on circles $p^2 + q^2 = constant$. Of course as $h \to 0$ the eccentricity of the ellipses decreases and they look more like circles: smaller values of h lead to smaller energy errors, as corresponds to a consistent method. Furthermore the fact that the computed points stay exactly on an ellipse near the theoretical circle implies that the error in energy remains bounded even if t becomes very large. □

Let us leave behind the class of linear Hamiltonian systems. For 'general' (autonomous) Hamiltonians, it is shown by Zhong and Marsden (1988) that a symplectic method $\psi_{h,H}$ *cannot* exactly conserve energy (except for the trivial cases where the function $\psi_{h,H}$ actually coincides with or is a time reparametrization of the true flow $\phi_{h,H}$). Hence conservation of the symplectic structure and conservation of energy are conflicting requirements that, in general, cannot be satisfied simultaneously by a numerical scheme. Since both the Hamiltonian and the symplectic structure are conserved by autonomous Hamiltonian systems, the question naturally arises of whether when constructing an integrator we should choose to conserve symplecticness and violate conservation of energy or vice versa. This is a question that should probably be answered differently for each specific application. However it should be pointed out that, as discussed in Chapter 2, symplecticness is a property that fully characterizes Hamiltonian problems, while conservation of an energy-like function is a feature also present in many non-Hamiltonian systems. Furthermore conservation of energy restricts the dynamics of the numerical solution by forcing the computed points to be on the correct $(2d - 1)$-dimensional manifold of constant H, but otherwise poses no restriction to the dynamics: within the manifold the points are free to move anywhere and only motions orthogonal to the manifold are forbidden. When d is large this is clearly a rather weak restriction. On the other hand, symplecticness restricts the dynamics in a more global way: all directions in phase space are taken into account.

Remark 10.1 The literature has devoted a great deal of attention to the construction of numerical schemes that exactly conserve H (or more generally, to the construction of integrators for, not neces-

CONSERVATION OF ENERGY

sarily Hamiltonian, differential systems that exactly conserve one or more invariants of motion). Several ideas have been used: (i) stepping from t_n to t_{n+1} with a standard method and then projecting the numerical result back onto the correct energy surface, (ii) adding the conservation constraints via Lagrange multipliers to the differential system (see the example below) to obtain a system of differential-algebraic equations (DAE's) that is integrated by a DAE solver (Brenan et al. (1989), Hairer et al. (1989), Hairer and Wanner (1991), Chapter VI), (iii) constructing *ad hoc* schemes. However conservation of energy is not the theme of this book and we shall not attempt to review the relevant literature. □

Example 10.4 For the pendulum system of Chapter 1

$$\dot{p} = -\sin q, \quad \dot{q} = p,$$

we consider the initial condition $p = 2$, $q = 0$, which lies on the separatrix

$$H = p^2/2 - \cos q = 1.$$

The augmented system

$$\dot{p} = -\sin q + \lambda p,$$
$$\dot{q} = p + \lambda \sin q,$$
$$p^2/2 - \cos q = 1,$$

where we have added $\lambda \nabla H$ to the right-hand side of the equation of motion (λ is the Lagrange multiplier), provides three differential-algebraic equations for the three unknown functions $p(t), q(t), \lambda(t)$. With initial conditions $p = 2$, $q = 0$, $\lambda = 0$, the augmented system has as solution the solution $(p(t), q(t))$ of the old pendulum problem along with $\lambda(t) \equiv 0$. The integration of the augmented system by the backward Euler rule leads to

$$p^{n+1} - p^n = h(-\sin q^{n+1} + \lambda^{n+1} p^{n+1}),$$
$$q^{n+1} - q^n = h(p^{n+1} + \lambda^{n+1} q^{n+1}),$$
$$(p^{n+1})^2/2 - \cos q^{n+1} = 1;$$

the last equation states that the numerical solution exactly conserves energy. Of course more sophisticated DAE integrators may be used. □

10.3.3 Approximate conservation

Even though, in general, symplectic integrators do not *exactly* conserve the value of the Hamiltonian H, they are very good at

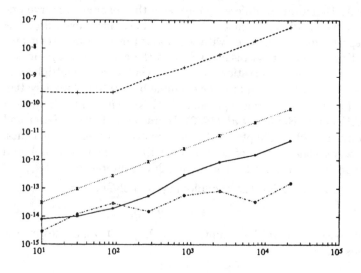

Figure 10.2. *Energy error as a function of t*

conserving it *approximately*. Figure 10.2 gives the error in H as a function of t (measured in periods) for the Kepler problem runs already considered in Fig. 9.3. In Fig. 9.3 we see that, at the final value of t, the four methods considered there produce errors in $(\mathbf{p}, \mathbf{q})$ in the range 2×10^{-2} to 2×10^{-3}. Fig. 10.2 reveals however that, in the same runs, the three symplectic algorithms yield *energy* errors three orders of magnitude smaller than the nonsymplectic algorithm.

The good (but approximate) conservation of H in symplectic integration is related to the existence of the modified Hamiltonian $\tilde{H}^h$ discussed after Example 10.1 above. The computed points do not remain exactly on trajectories of the modified problem $S_{\tilde{H}^h}$, but their drift away from the modified trajectories is very slow: the numerical scheme has local errors exponentially small when seen as approximations to $S_{\tilde{H}^h}$. Therefore the value of $\tilde{H}^h$ (which is constant along modified trajectories) is conserved by the numerical solutions, except for exponentially small errors, for long $O(1/h)$ time intervals. This in turn implies that the errors in $H = \tilde{H}^h + O(h^r)$ (r is the order of the method) possess an $O(h^r)$ bound on time intervals of length $O(1/h)$ (Lasagni (1988)). See also Calvo and Sanz-Serna (1993b) for the specific case of Kepler's problem.

10.4 KAM theory

The KAM theory, mentioned at the very end of Chapter 1, explains the behaviour of Hamiltonian systems that are perturbations of integrable systems. The theory also caters for symplectic mappings and can therefore be applied to investigate the properties of the mappings $\psi_{h,H}$ associated with symplectic integrators.

To get the flavour of this sort of application, let us consider once more the method (1)[1] in (10.2) applied to the system (10.1). Recall that the origin is a (stable) centre for this system. For the discrete equations (10.2), linearization around the origin results in

$$p^{n+1} = p^n + hf(0)q^{n+1}, \qquad q^{n+1} = q^n + hg(0)p^n, \qquad (10.9)$$

a transformation that has, for h small, unit modulus eigenvalues. Thus the origin is also a centre for (10.9). However to go from (10.9) to the full discretization (10.2) we must include the effects of the nonlinear terms that were discarded in the process of linearization. Since (10.9) is only neutrally stable, it may be feared that the nonlinear effects, small as they may be, will render the origin unstable for (10.2). In cases where the series H^h_∞ converges (see the discussion around Example 10.1), the computed points (p^n, q^n) lie exactly on trajectories of $S_{H^h_\infty}$ and hence stay on the level curves $H^h_\infty = constant$ in the phase plane: in these cases the computed points cannot leave the neighbourhood of the origin and it is clear that the nonlinear effects are *not* destabilizing. However we mentioned that for general nonlinear problems the series H^h_∞ does not converge and the computed points (p^n, q^n) do not stay on curves in the phase (p, q)-plane: the question of the stability of the origin arises again. KAM theory can be used to show that the symplecticness of the method implies that the destabilization by nonlinear effects does not actually take place. For full details see Sanz-Serna (1991).

CHAPTER 11

Generating functions

11.1 The concept of generating function

11.1.1 Introduction

We have now completed the 'basic' theory of symplectic integration. In order to make further progress, we need to present some tools of the Hamiltonian formalism that were not introduced in Chapters 1 and 2. We devote this chapter to one of those tools: the *generating function* of a symplectic transformation.

It is a remarkable feature of Hamiltonian problems that each system of the form (1.1) is fully determined by the choice of a *scalar* function H, whereas a general system of differential equations (3.1) is determined by a *vector field* $\mathbf{F}$. In a similar vein, a symplectic transformation $(\mathbf{p}^*, \mathbf{q}^*) = \psi(\mathbf{p}, \mathbf{q})$ can be expressed in terms of a single real-valued function S, rather than in terms of the $2d$ components of ψ. The function S is called the *generating function* of ψ.

11.1.2 Generating functions of the first kind

Let $(\mathbf{p}^*, \mathbf{q}^*) = \psi(\mathbf{p}, \mathbf{q})$ be a symplectic transformation defined in a simply connected domain Ω. For each closed path γ in Ω

$$\int_\gamma \mathbf{p}\,d\mathbf{q} - \int_\gamma \mathbf{p}^*\,d\mathbf{q}^* = 0, \tag{11.1}$$

where $\mathbf{p}\,d\mathbf{q}$ is the differential form $p_1 dq_1 + \cdots + p_d dq_d$, etc... In fact, by Stokes theorem, the first integral is the quantity $m(D)$, where m is the sum of two-dimensional areas considered in Section 2.5 and D is any two-dimensional surface bounded by γ. The second integral is $m(\psi(D))$ and hence (11.1) is just a way of saying that ψ is symplectic.

The key observation is that (11.1) is the condition for $\mathbf{p}\,d\mathbf{q}$ −

$\mathbf{p}^* d\mathbf{q}^*$ to be the differential of a function S defined in Ω:

$$dS = \mathbf{p}\, d\mathbf{q} - \mathbf{p}^* d\mathbf{q}^*. \tag{11.2}$$

Now let us further assume that $\mathbf{q}$ and $\mathbf{q}^*$ are *independent* functions in Ω, i.e., each point in Ω may be uniquely specified by the corresponding values of $\mathbf{q}$ and $\mathbf{q}^*$. Then we can express $S(\mathbf{p}, \mathbf{q})$ in (11.2) as a function S^1 of $\mathbf{q}$ and $\mathbf{q}^*$. It is evident from (11.2) that

$$\mathbf{p} = \frac{\partial S^1}{\partial \mathbf{q}}, \qquad \mathbf{p}^* = -\frac{\partial S^1}{\partial \mathbf{q}^*}. \tag{11.3}$$

These formulae implicitly define ψ by providing $2d$ relations among the $4d$ components of $\mathbf{q}, \mathbf{p}, \mathbf{p}^*, \mathbf{q}^*$. The function $S^1(\mathbf{q}, \mathbf{q}^*)$ is called the generating function (of the first kind) of ψ.

Example 11.1 The reader may wish to check that

$$S^1(q, q^*) = (\csc t)\, qq^* - \frac{\cot t}{2}\left(q^2 + q^{*2}\right) \tag{11.4}$$

is the generating function for the rotation given by (1.6) with $m = 1$, $\omega = 1$. □

Conversely, if we choose any smooth function $S^1(\mathbf{q}, \mathbf{q}^*)$ satisfying the condition that the Hessian determinant $\det(\partial^2 S^1/\partial \mathbf{q} \partial \mathbf{q}^*)$ does not vanish at a point $(\mathbf{q}_0, \mathbf{q}_0^*)$, then the formulae (11.3) implicitly define, in the neighbourhood of $(\mathbf{q}_0, \mathbf{q}_0^*)$, a *symplectic* transformation (see e.g. Arnold (1989), Section 47A).

11.1.3 Generating functions of the third kind

For a symplectic transformation ψ to have a generating function of the first kind, it is clearly necessary that $\mathbf{q}$ and $\mathbf{q}^*$ are *independent*, a condition not fulfilled by the identity transformation. (Note that (11.4) has a singularity at $t = 0$, where the rotation (1.6) is just the identity.) Since we are interested in generating consistent numerical methods Ψ, which, at $h = 0$, give the identity transformation, generating functions of the first kind are not really what we want.

Let us proceed as follows. Note that from (11.2)

$$d(\mathbf{p}^T \mathbf{q} - S) = \mathbf{q}\, d\mathbf{p} + \mathbf{p}^* d\mathbf{q}^* \tag{11.5}$$

and now assume that $\mathbf{p}$ and $\mathbf{q}^*$ are *independent* functions (which they are for the identity transformation). Then we can express the function in brackets in (11.5) in terms of the independent variables $\mathbf{p}$ and $\mathbf{q}^*$. The result $S^3(\mathbf{p}, \mathbf{q}^*)$ is called the generating function of the third kind of ψ and, from (11.5) we conclude that the formulae that now implicitly define ψ when S^3 is known are

$$\mathbf{p}^* = \frac{\partial S^3}{\partial \mathbf{q}^*}, \qquad \mathbf{q} = \frac{\partial S^3}{\partial \mathbf{p}}. \tag{11.6}$$

Example 11.2 The generating function of the identity is $\mathbf{p}^T \mathbf{q}^*$. For the rotation (1.6) ($m = \omega = 1$) we find

$$S^3(p, q^*) = (\sec t)\, pq^* - \frac{\tan t}{2}\left(p^2 + q^{*2}\right);$$

this is regular near $t = 0$, but breaks down when t approaches $\pm\pi/2$: at these values $p = q^*$ and p and q^* cannot be taken as independent coordinates. □

Conversely, given a function $S^3(\mathbf{p}, \mathbf{q}^*)$ with a locally nonvanishing Hessian det $\partial^2 S^3/\partial \mathbf{p}\partial \mathbf{q}^*$, the formulae (11.6) locally define a symplectic transformation.

11.1.4 Generating functions of all kinds

Some classical books on Hamiltonian mechanics considered four kinds of generating functions. Arnold (1989) has 2^n kinds. And in fact there are many more. The general idea is as follows (Feng (1986a), Feng and Qin (1987), Feng et al. (1989), Wu (1988), Feng and Qin (1991)). Collect in a ($2d$-dimensional) vector $\mathbf{y}$ the components of $(\mathbf{p}, \mathbf{q})$ and in a vector $\mathbf{y}^*$ the components of $(\mathbf{p}^*, \mathbf{q}^*)$. Introduce new $2d$-dimensional variables $\mathbf{w}$ and $\mathbf{w}^*$, such that $\mathbf{w}$ and $\mathbf{w}^*$ are linear functions of $\mathbf{y}$ and $\mathbf{y}^*$ (i.e. $\mathbf{w} = A\mathbf{y} + B\mathbf{y}^*$ and $\mathbf{w} = C\mathbf{y} + D\mathbf{y}^*$ for fixed $2d \times 2d$ matrices A, B, C, D). Under suitable hypotheses, the symplectic transformation $(\mathbf{p}^*, \mathbf{q}^*) = \psi(\mathbf{p}, \mathbf{q})$ reads, in terms of $\mathbf{w}$ and $\mathbf{w}^*$, $\mathbf{w}^* = \chi(\mathbf{w})$, where χ is the gradient of a scalar generating function σ. In the case of the generating functions of the first kind, A is the matrix that extracts the $\mathbf{q}$ variables of $\mathbf{y}$, B is the matrix that extracts the $\mathbf{q}^*$ variables of $\mathbf{y}^*$ etc. ...

A useful generating function is the so-called *Poincaré* generating function. Here $\mathbf{w}$ is taken to be the average of $\mathbf{y}$ and $\mathbf{y}^*$. The formulae for the transformation are (cf. MacKay (1992))

$$\mathbf{p}^* = \mathbf{p} - \partial_2 S^P\left(\frac{\mathbf{p}^* + \mathbf{p}}{2}, \frac{\mathbf{q}^* + \mathbf{q}}{2}\right),$$
$$\mathbf{q}^* = \mathbf{q} + \partial_1 S^P\left(\frac{\mathbf{p}^* + \mathbf{p}}{2}, \frac{\mathbf{q}^* + \mathbf{q}}{2}\right),$$

where ∂_1 and ∂_2 respectively represent differentiation with respect to the first and second groups of arguments in S^P.

Example 11.3 The Poincaré generating function of the identity is the 0 function. The function

$$\tan(t/2)\left[\left(\frac{\mathbf{p}^*+\mathbf{p}}{2}\right)^2+\left(\frac{\mathbf{q}^*+\mathbf{q}}{2}\right)^2\right]$$

generates the rotation considered in the examples above. This becomes singular at $t=\pi$, when the variables $(\mathbf{p}^*+\mathbf{p})/2$ and $(\mathbf{q}^*+\mathbf{q})/2$ vanish identically. □

11.2 Hamilton-Jacobi equations

Let us now consider symplectic transformations $(\mathbf{p}^*, \mathbf{q}^*) = \psi_t(\mathbf{p}, \mathbf{q})$ that depend on t as a parameter (an instance is given by the rotation in the examples of Section 11.1). We assume that ψ_t has a generating function of the third kind S^3, which of course depends on t. Let us further consider a Hamiltonian system (1.1) in the variables $\mathbf{p}, \mathbf{q}$. If we change variables in this system we obtain a new differential system for the new unknowns $\mathbf{p}^*, \mathbf{q}^*$. Then the following holds true (Arnold (1989), Section 45A).

Theorem 11.1 *In the situation above, the transformed system is also a Hamiltonian system, with the Hamiltonian function*

$$H^*(\mathbf{p}^*, \mathbf{q}^*, t) = H - \frac{\partial S^3}{\partial t}. \tag{11.7}$$

In (11.7) it is understood that once S^3 has been differentiated with respect to t with $\mathbf{p}$ and $\mathbf{q}^*$ constant, the formulae (11.6) that define the transformation are used to express the right-hand side in terms of the new variables $\mathbf{p}^*$ and $\mathbf{q}^*$.

A first corollary of this result refers to the case where the transformation is actually independent of t: then in the new variables the Hamiltonian system is still a Hamiltonian system and the new Hamiltonian is simply obtained by changing variables in the old Hamiltonian. We tacitly used this fact in Chapter 1: there we wrote the Kepler equations in polar coordinates by changing variables in H and then forming the associated Hamiltonian system S_H. The theorem guarantees that this is equivalent to changing coordinates in the equation of motion.

Another very important application of the theorem concerns the Hamilton-Jacobi equation that we study next. Assume that ψ_t is the exact flow $\phi_{t,H}$ of an autonomous Hamiltonian system

(1.1) and see the old variables $(\mathbf{p},\mathbf{q})$ evolving under the Hamiltonian system with Hamiltonian $-H(\mathbf{p},\mathbf{q},t)$. Changing sign in H is equivalent to reversing t, so that $(\mathbf{p},\mathbf{q})$ evolve under the operator ψ_{-t}. Then, the symplectic change of variables ψ_t just undoes what the Hamiltonian evolution under $-H$ does; in the new variables, the solutions of the differential equations are $\mathbf{p}^* =$ constant and $\mathbf{q}^* =$ constant and the new Hamiltonian $H^* = -H - \partial S^3/\partial t$ must be 0. We have proved that the generating function S^3 of the flow of the Hamiltonian system with Hamiltonian H satisfies

$$\frac{\partial S^3}{\partial t}(\mathbf{p},\mathbf{q}^*,t) + H(\mathbf{p},\mathbf{q},t) = 0. \tag{11.8}$$

This is the celebrated *Hamilton-Jacobi* equation. Upon replacing $\mathbf{q}$ by $\partial S^3/\partial \mathbf{p}$ (cf. (11.6)), the relation (11.8) is a partial differential equation of the first order for a function S^3 of the variables $\mathbf{p}$ and t (the $\mathbf{q}^*$'s act just as parameters). If this equation can be solved explicitly, we find the generating function of the flow and hence the solution of the system (1.1). This is Jacobi's approach to the solution of Hamilton's equation. Jacobi and others used this technique to explicitly integrate problems of mechanics that had proved intractable by other techniques (see e.g. Arnold (1989), Section 47). On the other hand, if we want to solve (11.8) by the method of characteristics, we find that the system of ordinary differential equations that defines the characteristics is none other than the system (1.1)! The equivalence between the solution of a Hamiltonian ordinary differential system and the solution of a first-order partial differential equation with Hamilton-Jacobi structure is thus complete.

The ideas above are not confined to generating functions of the third kind; they do work for all kinds of generating functions. The details of the construction of the new Hamiltonian H^* (and hence the form of the Hamilton-Jacobi equation) vary with the kind of generating function being used. The interested reader is referred to Feng (1986a), Feng and Qin (1987), Feng et al. (1989), Wu (1988), Feng and Qin (1991).

11.3 Symplectic integrators based on generating functions

The theorem in the preceding section is the key to the construction of symplectic integrators via generating functions, Channel (1983), Menyuk (1984), Feng (1986a), Feng and Qin (1987), Wu

(1988), Feng et al. (1989), Channell and Scovel (1990), Feng and Qin (1991), Miesbach and Pesch (1992). Let $\psi_{t,H}$ be a symplectic numerical method (H autonomous) of order r and with generating function S^3. An argument similar to that leading to the Hamilton-Jacobi equation, proves that $H^* = -H - \partial S^3/\partial t$ is $O(t^r)$ as $t \to 0$; now the transformation $\psi_{t,H}$ undoes the effect of the evolution under $-H$ except for terms of order $O(t^{r+1})$ and the corresponding Hamiltonian H^* must be $O(t^r)$. Conversely, any function S^3 that makes $H^* = O(t^r)$ generates a symplectic, r-th order numerical method, see Sanz-Serna and Abia (1991), Theorem 6.1.

Feng and his coworkers (Feng (1986a), Feng and Qin (1987), Feng et al. (1989), Wu (1988), Feng and Qin (1991)) take the following approach. They begin by expanding S^3 in (11.8) in powers of t. On substituting this power series in (11.8), expanding H and collecting like powers of t, the generating function S^3 can be expressed in terms of derivatives of H. When the series for S^3 is truncated, an approximate solution of the Hamilton-Jacobi is obtained, which is then used to generate the numerical method via (11.6).

Of course, similar approaches can be taken for generating functions other than generating functions of the third kind. The use of the Poincaré format is appealing, because it easily leads to *symmetric* schemes (see Section 3.6). The order 2 method derived from the Poincaré generating function is none other than the midpoint rule.

The expression for the fourth-order method turns out to be (Feng and Qin (1987))

$$p_i^{n+1} = p_i^n - hH_{q_i} - \frac{h^3}{24} \big[H_{p_j p_k q_i} H_{q_j} H_{q_k} + 2H_{p_j p_k} H_{q_j q_i} H_{q_k}$$
$$- 2H_{p_j q_k q_i} H_{p_j} H_{q_k} - 2H_{p_j q_k} H_{p_j q_i} H_{q_k} - 2H_{p_j q_k} H_{p_j} H_{q_k q_i}$$
$$+ 2H_{q_j q_k} H_{p_j q_i} H_{p_k} + H_{q_j q_k q_i} H_{p_j} H_{p_k} \big],$$

$$q_i^{n+1} = q_i^n + hH_{p_i} + \frac{h^3}{24} \big[H_{p_j p_k p_i} H_{q_j} H_{q_k} + 2H_{p_j p_k} H_{q_j p_i} H_{q_k}$$
$$- 2H_{p_j q_k p_i} H_{p_j} H_{q_k} - 2H_{p_j q_k} H_{p_j p_i} H_{q_k} - 2H_{p_j q_k} H_{p_j} H_{q_k p_i}$$
$$+ 2H_{q_j q_k} H_{p_j p_i} H_{p_k} + H_{q_j q_k p_i} H_{p_j} H_{p_k} \big].$$

Here summation in repeated indices must be understood and the functions featuring in the right-hand sides are evaluated at the averages

$$[(1/2)(\mathbf{p}^n + \mathbf{p}^{n+1}), (1/2)(\mathbf{q}^n + \mathbf{q}^{n+1})],$$

so that the scheme is implicit. We have reported these formulae to emphasize that methods of this sort, as distinct from RK, PRK or RKN methods, require higher derivatives of the Hamiltonian function. We point out that Feng's methods do not belong to the known class of higher derivative RK methods (Hairer *et al.* (1987), Chapter II, Section 12).

Miesbach and Pesch (1992) obtain methods that are based on generating functions and do not require higher derivatives of H.

11.4 Generating functions for symplectic Runga-Kutta methods

A symplectic Runge-Kutta method (3.11), (6.5) defines a symplectic transformation which, as the step size tends to 0, approaches the identity. Hence it must have, for h small, an S^3 generating function. Lasagni (1990) has found the corresponding expression

$$S^3(\mathbf{p}^n, \mathbf{q}^{n+1}, h) = \mathbf{p}^{nT}\mathbf{q}^{n+1} - h\sum_i b_i H(\mathbf{P}_i, \mathbf{Q}_i)$$
$$- h^2 \sum_{ij} b_i a_{ij} H_\mathbf{p}(\mathbf{P}_i, \mathbf{Q}_i) H_\mathbf{q}(\mathbf{P}_j, \mathbf{Q}_j)^T.$$

Here H is assumed autonomous (cf. Remark 4.1), $H_\mathbf{p}$ and $H_\mathbf{q}$ are row vectors of partial derivatives and of course the stages should be interpreted as functions of $\mathbf{p}^n$, $\mathbf{q}^{n+1}$ and the step size h implicitly defined in (6.1)–(6.4).

Remark 11.1 We showed above that a generating function would exist *if the domain Ω were simply connected.* Of course, Lasagni's recipe for S^3 works for all domains. Symplectic RK's have generating functions regardless of the geometry of Ω and therefore, in symplectic geometry jargon, they give rise to *exact symplectic* transformations, i.e. transformations for which (11.1) holds. Actually, the flow of a Hamiltonian system is also an exact symplectic transformation. □

In a similar manner for symplectic PRK, Abia and Sanz-Serna (1993) find the generating function

$$S^3(\mathbf{p}^n, \mathbf{q}^{n+1}, h) = \mathbf{p}^{nT}\mathbf{q}^{n+1} - h\sum_i b_i V(\mathbf{Q}_i)$$
$$- h\sum_i B_i T(\mathbf{P}_i) + h^2 \sum_{ij} B_i a_{ij} \mathbf{g}(\mathbf{P}_i)^T \mathbf{f}(\mathbf{Q}_j)$$

and for symplectic RKN schemes the generating function is given by (Calvo and Sanz-Serna (1992a))

$$S^3(\mathbf{p}^n, \mathbf{q}^{n+1}, h) = \mathbf{p}^{nT}\mathbf{q}^{n+1} - h\sum_i b_i V(\mathbf{Q}_i)\frac{h}{2}\mathbf{p}^{nT}\mathbf{p}^n$$
$$-+\frac{h^3}{2}\sum_{ij} b_i(\beta_j - \alpha_{ij})\,\mathbf{f}(\mathbf{Q}_i)^T\mathbf{f}(\mathbf{Q}_j).$$

We emphasize that, unlike the situation with the methods considered in the preceding section, these generating functions are not needed to derive or to implement the corresponding RK, PRK or RKN method. The explicit knowledge of the generating functions may, however, be put to good use in the study of the order conditions. This is shown in the next section.

11.5 The canonical theory of the order. Elementary Hamiltonians

11.5.1 General framework

In the discussion at the beginning of Section 11.2, we pointed out a canonical numerical method $\psi_{h,H}$ (H autonomous, cf. Remark 4.1) is of order r if and only if

$$H^* = H + \partial S^3/\partial h = O(h^r), \quad h \to 0, \qquad (11.9)$$

where $S^3(\mathbf{p}^n, \mathbf{q}^{n+1}, h)$ denotes the corresponding generating function.

Of course we impose the requirement (11.9) by demanding that at $h = 0$, the partial derivatives of order $0, 1, \ldots, r-1$ of H^* with respect to h vanish. These partial derivatives are derivatives with respect to t with the other arguments of H^*, namely $\mathbf{p}^{n+1}$ and $\mathbf{q}^{n+1}$, held fixed. However, if the method is consistent $(\mathbf{p}^{n+1}, \mathbf{q}^{n+1})$ and $(\mathbf{p}^n, \mathbf{q}^n)$ differ in $O(h)$ terms and the derivatives with respect to h with $(\mathbf{p}^{n+1}, \mathbf{q}^{n+1})$ fixed vanish if and only if the derivatives with respect to h with $(\mathbf{p}^n, \mathbf{q}^n)$ fixed vanish.

These remarks lead to the following technique for investigating the order of a symplectic method $\psi_{h,H}$. We begin by finding the generating function $S^3(\mathbf{p}^n, \mathbf{q}^{n+1}, h)$. Then we differentiate with respect to h with $\mathbf{p}^n$ and $\mathbf{q}^{n+1}$ fixed and in the result, we write $\mathbf{q}^{n+1}$ in terms of $\mathbf{p}^n$, $\mathbf{q}^n$ and h. This yields a function $\Gamma = \Gamma(\mathbf{p}^n, \mathbf{q}^n, h)$. Then, assuming order $\geq r - 1$, with $r \geq 2$, for order $r \geq 2$ we

CANONICAL ORDER THEORY

demand that, at $h = 0$,

$$\frac{\partial^{r-1}\Gamma}{\partial h^{r-1}} = 0. \tag{11.10}$$

This is the methodology for investigating the order suggested by Sanz-Serna and Abia (1991). We illustrate it below for the case of canonical PRK methods (Abia and Sanz-Serna (1993)). For the RK case see Sanz-Serna and Abia (1991) and Abia and Sanz-Serna (1993). Calvo (1992) has studied the RKN case.

11.5.2 The Partitioned Runge-Kutta case

With the expression for S^3 given in Section 11.4 straightforward computation reveals that, for a symplectic PRK method applied to the separable Hamiltonian system (6.8),

$$\Gamma = -\sum_i b_i V(\mathbf{Q}_i) - \sum_i B_i T(\mathbf{P}_i), \tag{11.11}$$

where, as usual, $\mathbf{P}_i$ and $\mathbf{Q}_i$ are the stage vectors for the p and q variables. The partial derivatives in (11.10) may be computed explicitly by using RK techniques. The interested reader is directed to Abia and Sanz-Serna (1993) for the details. The result is

$$\frac{\partial^{r-1}\Gamma}{\partial h^{r-1}} = -\frac{1}{r} \sum_{\beta\rho\tau \text{ has order } r} \alpha(\beta\rho\tau)\,\gamma(\beta\rho\tau)\,\Phi(\beta\rho\tau)\,\theta(\beta\rho\tau), \tag{11.12}$$

where $\theta(\beta\rho\tau)$ is a real-valued function of $\mathbf{p}^n$ and $\mathbf{q}^n$, whose construction is described next. For the bicolour rooted trees with three vertices (see Fig. 4.3), the explicit expressions are (superscripts denote components)

$$\theta(\beta\rho\tau_{3,1,b}) = \sum_{IJ} \frac{\partial T}{\partial p^I} \frac{\partial f^I}{\partial q^J} g^J,$$

$$\theta(\beta\rho\tau_{3,1,w}) = \sum_{IJ} \frac{\partial V}{\partial q^I} \frac{\partial g^I}{\partial p^J} f^J,$$

$$\theta(\beta\rho\tau_{3,2,b}) = \sum_{IJ} \frac{\partial^2 T}{\partial p^I \partial p^J} f^I f^J,$$

$$\theta(\beta\rho\tau_{3,2,w}) = \sum_{IJ} \frac{\partial^2 V}{\partial q^I \partial q^J} g^I g^J.$$

In general, there are as many summation indices as vertices other than the root. A white (resp. black) root with ℓ sons $I, J, \ldots$ introduces a factor $\partial^\ell V/\partial q^I \partial q^J \ldots$ (resp. $\partial^\ell T/\partial p^I \partial p^J \ldots$). A white (resp. black) vertex, different from the root, with label I and ℓ sons labelled $J, K, \ldots$ brings in a factor $\partial^\ell f^I/\partial q^J \partial q^K \ldots$ (resp. $\partial^\ell g^I/\partial p^J \partial p^K \ldots$).

In the expression for θ, we write $\mathbf{f}$ in terms of V and $\mathbf{g}$ in terms of T. For the bicolour rooted trees with three vertices the result is (subscripts denote derivatives)

$$\theta(\beta\rho\tau_{3,1,b}) = -\sum_{IJ} T_I V_{IJ} T_J,$$

$$\theta(\beta\rho\tau_{3,1,w}) = -\sum_{IJ} V_I T_{IJ} V_J,$$

$$\theta(\beta\rho\tau_{3,2,b}) = \sum_{IJ} T_{IJ} V_I V_J,$$

$$\theta(\beta\rho\tau_{3,2,w}) = \sum_{IJ} V_{IJ} T_I T_J.$$

We see that the first two sums are preceded by a $-$ sign. This is because the definition of the corresponding θ had an odd number of factors f and such a factor introduces a change in sign when written in terms of V. Next observe that the first and fourth sums are identical, and so are the second and third sums. This is because $\beta\rho\tau_{3,1,b}$ and $\beta\rho\tau_{3,2,w}$ are representatives of the same (unrooted) bicolour tree $\beta\tau_{3,1}$; $\beta\rho\tau_{3,1,w}$ and $\beta\rho\tau_{3,2,b}$ are representatives of $\beta\tau_{3,2}$. In the general case we find that

$$\theta(\beta\rho\tau) = (-1)^{w(\beta\rho\tau)} \mathcal{H}(\beta\tau),$$

where $w(\beta\rho\tau)$ is the number of white vertices other than the root in $\beta\rho\tau$ and $\mathcal{H}(\beta\tau)$ depends only on the bicolour tree associated with $\beta\rho\tau$. The expression $\mathcal{H}(\beta\tau)$ is a summation over as many indices as *edges* there are in $\beta\tau$. A white (resp. black) vertex entered by edges labelled $I, J, \ldots$ introduces a factor $V_{IJ\ldots}$ (resp. $T_{IJ\ldots}$).

Taking these considerations into account, (11.12) may be rewritten as

$$\frac{\partial^{r-1}\Gamma}{\partial h^{r-1}} = -\frac{1}{r} \sum_{\beta\tau \text{ order } r} (-1)^{w(\beta\tau)+1}$$
$$\left[\sum_{\substack{\beta\rho\tau \in \beta\tau \\ \text{white root}}} \alpha(\beta\rho\tau)\, \gamma(\beta\rho\tau)\, \Phi(\beta\rho\tau) \right.$$

$$-\sum_{\substack{\beta\rho\tau \in \beta\tau \\ \text{black root}}} \alpha(\beta\rho\tau)\,\gamma(\beta\rho\tau)\,\Phi(\beta\rho\tau) \Bigg]\mathcal{H}(\beta\tau), \quad (11.13)$$

where $w(\beta\tau)$ denotes the number of white vertices in $\beta\tau$. This construction explains why the homogeneous form of the order conditions (7.13) works: the homogeneous form demands that the term in brackets in (11.13) vanishes for each $\beta\rho$ of order r. Then, because of (11.10), it is clear that the equations (7.13) are *sufficient* to ensure order r. They are also necessary because the $\mathcal{H}$ corresponding to different bicolour trees are independent (Abia and Sanz-Serna (1993), Theorem 4.2) and hence (11.13) cannot vanish unless the term in brackets is 0 for each $\beta\rho$ of order r.

11.5.3 Elementary Hamiltonians

The functions $\mathcal{H}(\beta\tau)$ were introduced by E. Hairer (1993), who calls them *elementary Hamiltonians*. They play, for symplectic integration, the role that for general integrators is played by the *elementary differentials*. Abia and Sanz-Serna (1993) work only in terms of the closely related θ functions, which they refer to as canonical elementary differentials. The terminology elementary Hamiltonian is to be preferred, in view that $\mathcal{H}$ and θ are scalar quantities.

The elementary Hamiltonians often appear in the theory of symplectic integration.

Example 11.4 For the unique two-vertex bicolour tree $\beta_{2,1}$ the elementary Hamiltonian is $\sum V_I T_I$. For a one-degree-of-freedom case this reduces to $V'T'$, the function that features in the modified Hamiltonian H_2 for the system (10.4). Similarly in the modified Hamiltonian H_3 in (10.5), we recognize in the functions $T''(V')^2$ and $V''(T')^2$ the elementary Hamiltonians for $\beta\rho_{3,2}$ and $\beta\rho_{3,1}$ respectively.

The expansion in powers of h of the Hamiltonian $\hat{H}$ in Example 10.2 also gives rise to elementary Hamiltonians. □

CHAPTER 12

Lie formalism

12.1 The Poisson bracket

We now turn our attention to the Poisson bracket, another important tool in the Hamiltonian formalism.

If F and G are smooth real functions defined in the phase space Ω (see Section 1.1) of the variables $(p_1, \ldots, p_d, q_1, \ldots, q_d)$, their *Poisson bracket* (Arnold (1989) Section 39) is defined as

$$\{F, G\} = \sum_{i=1}^{d} \left(\frac{\partial F}{\partial q_i} \frac{\partial G}{\partial p_i} - \frac{\partial F}{\partial p_i} \frac{\partial G}{\partial q_i} \right). \tag{12.1}$$

The following properties are easily checked.

(i) *Bilinearity* For constants λ_1, λ_2 and real functions F, F_1, F_2, G, G_1, G_2

$$\{\lambda_1 F_1 + \lambda_2 F_2, G\} = \{\lambda_1 F_1, G\} + \{\lambda_2 F_2, G\},$$
$$\{F, \lambda_1 G_1 + \lambda_2 G_2\} = \{F, \lambda_1 G_1\} + \{F, \lambda_2 G_2\}.$$

(ii) *Skew-symmetry* $\{F, G\} = -\{G, F\}$.

(iii) *Jacobi condition*

$$\{F, \{G, H\}\} + \{G, \{H, F\}\} + \{H, \{F, G\}\} = 0.$$

Due to these properties, the vector space $C^\infty(\Omega)$ of all indefinitely differentiable real functions defined in Ω, endowed with the product operation $\{\cdot, \cdot\}$, is a *Lie algebra*. A further property is:

(iv) *Leibniz' rule*

$$\{F, G \cdot H\} = \{F, G\} \cdot H + G \cdot \{F, H\}.$$

Here $\cdot$ denotes ordinary multiplication of real-valued functions.

Upon introducing a $2d$-dimensional vector $\mathbf{y} = (\mathbf{p}, \mathbf{q})$ as in Section 1.1, the definiton in (12.1) becomes

$$\{F, G\} = \nabla F^T J^{-1} \nabla G, \tag{12.2}$$

where J is the skew-symmetric matrix (1.4) and ∇ is the gradient operator (1.3).

The Poisson bracket may be used to express the basic relations of the Hamiltonian formalism. For instance, in view of (12.2), the equations of motion (1.2) can be written as

$$\dot{y}_i = \{y_i, H\}, \qquad i = 1, \ldots, 2d. \tag{12.3}$$

Furthermore, the condition (2.6) for a transformation $\mathbf{y} \to \mathbf{y}^* = \psi(\mathbf{y})$ to be canonical, becomes (after noticing that $J^{-1} = J^T = -J$)

$$\{y_i^*, y_j^*\} = -J_{ij}, \qquad i,j = 1, \ldots, 2d. \tag{12.4}$$

Remark 12.1 Note that, if we saw $\mathbf{y}^*$ as an independent variable ranging in the image $\Omega^* = \psi(\Omega)$, then

$$\{y_i^*, y_j^*\}_{\Omega^*} = (\nabla_\mathbf{y} \cdot y_i^*)^T J^{-1} (\nabla_\mathbf{y} \cdot y_j^*) = -J_{ij}.$$

Hence (12.4) may be rephrased by saying that the transformation is canonical if and only if the Poisson bracket $\{y_i^*, y_j^*\}$ is the same whether one sees $\mathbf{y}^*$ as a function of $\mathbf{y}$ or as the independent variable. From here, it follows easily that for any two functions $F(\mathbf{y}^*)$, $G(\mathbf{y}^*)$

$$\{F(\mathbf{y}^*), G(\mathbf{y}^*)\}_{\Omega^*} = \{F(\mathbf{y}^*(\mathbf{y})), G(\mathbf{y}^*(\mathbf{y}))\}_\Omega.$$

To sum up: $\mathbf{y} \to \mathbf{y}^*$ *is canonical if and only if the Poisson bracket of any two functions remains invariant in the change of variables.* □

12.2 Lie operators and Lie series

12.2.1 Lie operators

With each function $F \in \mathcal{C}^\infty(\Omega)$ we associate a *Lie operator* $\widehat{F}$ (see e.g. Dragt and Finn (1976)). By definition $\widehat{F}$ is the operator that transforms each function $G \in \mathcal{C}^\infty(\Omega)$ into the function $\widehat{F}G = \{F, G\}$. According to (12.2) we can write $\widehat{F} = \nabla F^T J^{-1} \nabla$, so that $\widehat{F}$ is a *first-order differential operator* in Ω.

In connection with Hamiltonian systems the meaning of $\widehat{F}$ is as follows. Consider the autonomous Hamiltonian system $\mathcal{S}_{-F}$ with Hamiltonian function $-F$ and let $\mathbf{y}(t)$ be a solution of $\mathcal{S}_{-F}$. Then for any function $G \in \mathcal{C}^\infty(\Omega)$

$$\begin{aligned} \frac{d}{dt} G(\mathbf{y}(t)) &= \nabla G^T \dot{\mathbf{y}} = \nabla G^T J^{-1} \nabla(-F) \\ &= \{G, -F\} = (\widehat{F}G)(\mathbf{y}(t)); \end{aligned} \tag{12.5}$$

LIE OPERATORS AND LIE SERIES

$\widehat{F}G$ measures the rate of change of G along solutions of S_{-F}. In particular $\widehat{F}G = 0$ or $\{F, G\} = 0$, if G is a conserved quantity of S_{-F} (or equivalently of S_F). By the skew-symmetry of the Poisson bracket, we conclude that G is a conserved quantity of S_F if and only if F is a conserved quantity of S_G.

It is possible to write (12.5) more elegantly in terms of flows:

$$\frac{d}{dt}G(\phi_{t,-F}(\mathbf{y}^0))\bigg|_{t=0} = (\widehat{F}G)(\mathbf{y}^0). \qquad (12.6)$$

12.2.2 The adjoint representation

Differential operators may be composed with each other in a standard way. In general they do not commute: if $\mathcal{F}$ and $\mathcal{G}$ are differential operators $\mathcal{F}\mathcal{G} \neq \mathcal{G}\mathcal{F}$ and the degree of noncommutativity is measured by the *commutator*

$$[\mathcal{F}, \mathcal{G}] = \mathcal{F}\mathcal{G} - \mathcal{G}\mathcal{F}.$$

If $\mathcal{F}$ and $\mathcal{G}$ are differential operators of order 1, then $\mathcal{F}\mathcal{G}$ and $\mathcal{G}\mathcal{F}$ are of order 2 (i.e., $\mathcal{F}\mathcal{G}G$ and $\mathcal{G}\mathcal{F}G$ involve second derivatives of G). However $[\mathcal{F}, \mathcal{G}]$ turns out to be again a first-order operator. In actual fact, if $\mathcal{F}$ and $\mathcal{G}$ are Lie operators respectively arising from functions F and G, i.e. $\mathcal{F} = \widehat{F}$, $\mathcal{G} = \widehat{G}$, then $[\mathcal{F}, \mathcal{G}]$ is the Lie operator $\widehat{\{F, G\}}$ corresponding to the function $\{F, G\}$, Poisson bracket of F and G. This follows from the Jacobi property of $\{\cdot, \cdot\}$. Thus the transformation $F \to \widehat{F}$, mapping functions into differential operators, is a homomorphism of the Lie algebra $C^\infty(\Omega)$ with the product $\{\cdot, \cdot\}$ into the Lie algebra of differential operators with the product $[\cdot, \cdot]$. This homomorphism is called *the adjoint representation*.

12.2.3 Lie series

It is possible to consider the powers $\widehat{F}^2, \widehat{F}^3, \ldots$ of a Lie operator $\widehat{f}$. These are of course defined by the standard recipe for composing operators

$$\widehat{F}^2 G = \widehat{F}(\widehat{F}(G)), \quad \ldots, \quad \widehat{F}^k G = \widehat{F}(\widehat{F}^{k-1}(G)).$$

Two applications of (12.5) reveal that

$$\frac{d^2}{dt^2}G(\mathbf{y}(t)) = \frac{d}{dt}\left[\frac{d}{dt}G(\mathbf{y}(t))\right]$$

$$= \frac{d}{dt}\left[(\widehat{F}G)(\mathbf{y}(t))\right] = [\widehat{F}(\widehat{F}G)](\mathbf{y}(t)),$$

or, in terms of flows,

$$\left.\frac{d^2}{dt^2}G(\phi_{t,-F}(\mathbf{y}^0))\right|_{t=0} = (\widehat{F}^2 G)(\mathbf{y}^0).$$

Similarly, for $k = 1, 2, 3, \ldots,$

$$\left.\frac{d^k}{dt^k}G(\phi_{t,-F}(\mathbf{y}^0))\right|_{t=0} = (\widehat{F}^k G)(\mathbf{y}^0); \qquad (12.7)$$

$\widehat{F}^k G$ thus provide the higher derivatives of G with respect to t along solutions of $\mathcal{S}_{-F}$.

As a further step, we may write the exponential $\exp(\widehat{F})$ of a Lie operator. We define $\exp(t\widehat{F})$ as the formal series of operators

$$\exp(t\widehat{F}) = I + t\widehat{F} + \frac{t^2}{2!}\widehat{F}^2 + \cdots + \frac{t^k}{k!}\widehat{F}^k + \cdots$$

Formal means that the series is seen a sequence of symbols without considering the question of convergence.

By (12.7),

$$[\exp(t\widehat{F})G](\mathbf{y}^0) = \sum_{k=0}^{\infty} \frac{t^k}{k!} \frac{d^k}{dt^k} G(\phi_{t,-F}(\mathbf{y}^0))\bigg|_{t=0}. \qquad (12.8)$$

The right-hand side of (12.8) is of course the Taylor series of $G(\phi_{t,-F}(\mathbf{y}^0))$, so that formally

$$[\exp(t\widehat{F})G](\cdot) = G(\phi_{t,-F}(\cdot)). \qquad (12.9)$$

Here the place-holder notation $\cdot$ is used to refer to the point in phase space at which the functions are evaluated.

The formula (12.9) provides a way for finding (the Taylor series of) the flow $\phi_{t,-F}$ (i.e. for integrating $\mathcal{S}_{-F}$) in terms of Lie operators. By successively choosing G to be one of the *coordinate functions*

$$\mathbf{y} = (y_1, \ldots, y_{2d}) \to y_i,$$

we see in (12.9) that $\exp(t\widehat{F})y_i$ is the i-th component of the flow. We emphasize that in $\exp(t\widehat{F})y_i$, the symbol y_i refers to a *function* on which the formal operator $\exp(t\widehat{F})$ acts to yield a new function. Interpreting y_i as the numerical value of the i-th component of a point in phase space is meaningless: Lie *operators* do not act on numbers. The following example may be helpful.

LIE OPERATORS AND LIE SERIES

Example 12.1 Assume that $d = 1$, so that $y = (p, q)$. Consider the Hamiltonian system $\dot p = -V'(q)$, $\dot q = 0$ with (separable) Hamiltonian $V(q)$. The system is trivially solved: q remains constant $q = q^0$ and p increases with constant velocity $p = p^0 - tV'(q^0)$.

Here, $-\widehat V$ is the differential operator that associates with each function $G(p, q)$ the function $-\widehat V G(p, q) = -\partial G/\partial p\, V'$. By iteration,
$$\left((-\widehat V)^k G\right)(p, q) = (-1)^k \frac{\partial^k G}{\partial p^k} V^{(k)}.$$
If $G(p, q) = p$ (first coordinate function) then $\partial G/\partial p = 1$, and, for $k \geq 2$, $\partial^k G/\partial p^k = 0$; therefore
$$(\exp[-t\widehat V]G)(p^0, q^0) = p^0 - tV'(q^0).$$
If $G(p, q) = q$ (second coordinate function) then $\partial^k G/\partial p^k = 0$ for $k \geq 1$ and
$$(\exp[-t\widehat V]G)(p^0, q^0) = q^0.$$
In $p^0 - tV'(q^0)$, q^0 we recognize the components of the solution (flow) of $\dot p = -V'(q)$, $\dot q = 0$. □

The reader may find the Lie series corresponding to the function $(1/2)(p^2 + q^2)$ in order to integrate the harmonic oscillator.

12.2.4 Multiplication of exponentials

What happens if we compose two exponential operators $\exp(t\widehat F_1)\exp(t\widehat F_2)$? The product represents the composition of flows $\phi_{t,-F_1}$ $\phi_{t,-F_2}$; more precisely
$$\left[\exp(t\widehat F_1)\exp(t\widehat F_2)\right]G(\cdot) = G(\phi_{t,-F_2}(\phi_{t,-F_1}(\cdot))). \qquad (12.10)$$
Some people may have expected that the recipe would have been
$$\left[\exp(t\widehat F_1)\exp(t\widehat F_2)G\right](\cdot) = G(\phi_{t,-F_1}(\phi_{t,-F_2}(\cdot))).$$
This expectation is wrong: it is based on the idea that Lie operators act on points and hence their composition behaves like the composition of functions.

Let us prove (12.10). We introduce the auxiliary function
$$\bar G = \exp(t\widehat F_2)G,$$
then, by (12.9),
$$\left[\exp(t\widehat F_1)\exp(t\widehat F_2)G\right](\cdot) = \left[\exp(t\widehat F_1)\bar G\right](\cdot) = \bar G(\phi_{t,-F_1}(\cdot)).$$

Replacement of $\bar{G}$ by its definition shows that
$$\left[\exp(t\widehat{F}_1)\exp(t\widehat{F}_2)G\right](\cdot) = \left[\exp(t\widehat{F}_2)G\right](\phi_{t,-F_1}(\cdot))$$
and now a new application of (12.9) leads to (12.10).

12.3 The Baker-Campbell-Hausdorff formula

We now study how to write the product of two exponentials as a new exponential. This will be very useful later.

Let X and Y be 'symbols'. (We later think of these as differential operators, but this is not necessary for the present.) We form the exponentials
$$\exp(X) = I + X + \frac{1}{2}X^2 + \frac{1}{6}X^3 + \cdots$$
$$\exp(Y) = I + Y + \frac{1}{2}Y^2 + \frac{1}{6}Y^3 + \cdots$$
and we multiply them out
$$\exp(X)\exp(Y) = I + X + Y + \frac{1}{2}X^2 + XY + \frac{1}{2}Y^2$$
$$+ \frac{1}{6}X^3 + \frac{1}{2}X^2Y + \frac{1}{2}XY^2 + \frac{1}{6}Y^3 + \cdots$$
According to the Baker-Campbell-Hausdorff (BCH) formula (Varadarajan (1974), Dragt and Finn (1976), Yoshida (1990)), the product $\exp(X)\exp(Y)$ can be written as the exponential $\exp(Z)$ of a new symbol
$$Z = X + Y + \frac{1}{2}[X,Y] + \frac{1}{12}([X,X,Y] + [Y,Y,X])$$
$$+ \frac{1}{24}[X,Y,Y,X] - \frac{1}{720}([Y,Y,Y,Y,X] + [X,X,X,X,Y])$$
$$+ \frac{1}{360}([Y,X,X,X,Y] + [X,Y,Y,Y,X])$$
$$+ \frac{1}{120}([X,X,Y,Y,X] + [Y,Y,X,X,Y]) + \cdots \quad (12.11)$$
Here $[X,Y]$ is the commutator $[X,Y] = XY - YX$ and we use iterated commutators like
$$[X,X,Y] = [X,[X,Y]]$$
$$[X,Y,Y,X] = [X,[Y,[Y,X]]],$$
etc. It is remarkable that Z consists of X, Y and commutators. In particular, if X and Y are Lie operators $\widehat{F}$, $\widehat{G}$, then Z is a new

Lie operator: the operator associated with the function $F + G + (1/2)\{F, G\} + \cdots$ This is made up of iterated Poisson brackets of F and G.

Two applications of (12.11) lead to the conclusion

$$\exp(X)\exp(Y)\exp(X) = \exp(W), \qquad (12.12)$$

with (Yoshida (1990))

$$\begin{aligned} W = {} & 2X + Y + \frac{1}{6}[Y,Y,X] - \frac{1}{6}[X,X,Y] \\ & + \frac{7}{360}[X,X,X,X,Y] - \frac{1}{360}[Y,Y,Y,Y,X] \\ & + \frac{1}{90}[X,Y,Y,Y,X] + \frac{1}{45}[Y,X,X,X,Y] \\ & - \frac{1}{60}[X,X,Y,Y,X] + \frac{1}{30}[Y,Y,X,X,Y] + \cdots \end{aligned}$$

12.4 Application to fractional-step methods

12.4.1 Introduction

The Lie formalism presented so far is the key of many developments in symplectic integration. Some of these developments will be presented in Chapter 13. We here consider the application to fractional-step methods.

Often the (autonomous) Hamiltonian H of the system S_H to be integrated can be written as $H = H_1 + H_2$ with the Hamiltonians H_1, H_2 'simpler' in some sense than H. It is then natural to look for numerical methods that advance the solutions of S_H by combining *fractional steps*, i.e. steps of a numerical method for S_{H_1} with steps of a numerical method for S_{H_2}.

Example 12.2 Consider a separable Hamiltonian $H = T(p) + V(q)$ (we assume, only for simplicity, that the number of degrees of freedom is 1). If we split H as $H = H_1 + H_2$, $H_1 = T$, $H_2 = V$, each of the systems S_{H_1}, S_{H_2} can be integrated exactly (cf. Example 12.1). In S_T p is constant and q varies linearly; in S_V q is constant and p varies linearly. We now replace the true evolution of S_H, where H_1 and H_2 contribute *simultaneously* by an evolution where H_1, H_2 act *successively*. Starting from (p^n, q^n) the evolution of h units of time with H_1 leads to

$$(\bar{p}, \bar{q}) = (p^n, q^n + hT'(p^n)).$$

Now evolving from $(\bar{p}, \bar{q})$ with H_2 results in
$$(p^{n+1}, q^{n+1}) = (p^n - hV'(\bar{q}), \bar{q}).$$
Elimination of $(\bar{p}, \bar{q})$ reveals that
$$p^{n+1} = p^n - hV'(q^{n+1}), \qquad q^{n+1} = q^n + hT'(p^n),$$
i.e. we are dealing with the (1)[1] method of (10.2). The overall method, the result of composing the two fractional steps, is given by
$$\psi_{h,H} = \phi_{h,H_2}\phi_{h,H_1}.$$
Now ϕ_{h,H_1} corresponds as in (12.9) to the exponential $\exp[-h\widehat{T}]$ and ϕ_{h,H_2} corresponds to $\exp[-h\widehat{V}]$. By the BCH formula (12.11), the product $\exp[-h\widehat{T}]\exp[-h\widehat{V}]$ equals $\exp[-h\widehat{H}^h_\infty]$ where H^h_∞ is the function
$$H^h_\infty(p,q) = T(p) + V(q) - \frac{h}{2}\{T,V\} + \frac{h^2}{12}(\{T,T,V\} + \{V,V,T\}) + \cdots$$
or, computing explicitly the Poisson brackets,
$$\begin{aligned}H^h_\infty(p,q) &= T(p) + V(q) + \frac{h}{2}T'(p)V'(q) \\ &\quad + \frac{h^2}{12}[T'(p)^2 V''(q) + V'(q)^2 T''(p)] + \cdots\end{aligned}$$
The h-flow of this Hamiltonian corresponds (see (12.10)) to the composition $\phi_{h,H_2}\phi_{h,H_1}$, i.e. to the mapping $\psi_{h,H}$. We have therefore found the expression for the (formal power series for the) *modified Hamiltonian* H^h_∞ for the method; the expression of course coincides with that found in (10.5) . □

In the example above the fractional systems S_{H_1}, S_{H_2} could be integrated exactly. This is not essential to construct a fractional-step method. If ϕ_{h,H_1}, ϕ_{h,H_2} are not available explicitly, one may approximate them by numerical methods $\psi^{[1]}_{h,H_1}$, $\psi^{[2]}_{h,H_2}$ and take $\psi_{h,H} = \psi^{[2]}_{h,H_2}\psi^{[1]}_{h,H_1}$ as a method for S_H. All that is required is that S_{H_1}, S_{H_2} be simpler than S_H. For instance, if S_H originates from the space discretization by finite differences or finite elements of a system of partial differential equations in two space dimensions, H_1, H_2 could represent the contributions in the x and y directions (dimensional splitting, see e.g. Strang (1963)). Then S_{H_1}, S_{H_2} are spatial discretizations of problems in one space dimension. Alternatively H_1 and H_2 may correspond to stiff and nonstiff parts, linear and nonlinear terms, etc. The situation where H_2 is a small

APPLICATION TO FRACTIONAL-STEP METHODS

perturbation of a Hamiltonian H_1 integrable in closed form is also important.

12.4.2 The simplest splitting

Assume that $\psi^{[1]}_{h,H_1}$, $\psi^{[2]}_{h,H_2}$ are consistent, symplectic methods for S_{H_1}, S_{H_2}. Then the simplest fractional-step method for $S_{H_1+H_2}$ is, as mentioned before,

$$\psi_{h,H} = \psi^{[2]}_{h,H_2}\psi^{[1]}_{h,H_1}. \tag{12.13}$$

This method is symplectic: the composition of symplectic maps is again symplectic. How about its accuracy? The methods $\psi^{[i]}_{h,H_i}$, $i = 1,2$, possess modified Hamiltonians $H^h_{\infty,i}$ with $\psi^{[i]}_{h,H_i} = \phi_{h,H^h_{\infty,i}}$. If both are of order r, then $H^h_{\infty,i} = H_i + O(h^r)$. By the BCH formula the modified Hamiltonian for $\psi_{h,H}$ is

$$H^h_\infty = H^h_{\infty,1} + H^h_{\infty,2} - \frac{h}{2}\{H^h_{\infty,1}, H^h_{\infty,2}\} \tag{12.14}$$
$$+ \frac{h^2}{12}(\{H^h_{\infty,1}, H^h_{\infty,1}, H^h_{\infty,2}\} + \{H^h_{\infty,2}, H^h_{\infty,2}, H^h_{\infty,1}\}) + \cdots$$

In the unlikely event that $\{H^h_{\infty,1}, H^h_{\infty,2}\} = 0$ all the Poisson brackets in (12.14) vanish and

$$H^h_\infty = H^h_{\infty,1} + H^h_{\infty,2} = H_1 + H_2 + O(h^r) = H + O(h^r).$$

The overall method is thus consistent of order r with H. However, in general $\{H^h_{\infty,1}, H^h_{\infty,2}\} \neq 0$ and $H^h_\infty = H + O(h)$ so that (12.13) is only of the *first order*, regardless of the value of r.

12.4.3 Second-order splitting

Better splittings than (12.13) are possible, for instance (Strang (1968))

$$\psi_{h,H} = \psi^{[1]}_{h/2,H_1}\psi^{[2]}_{h,H_2}\psi^{[1]}_{h/2,H_1} \tag{12.15}$$

is a popular choice. Now according to (12.12)

$$H^h_\infty = H^h_{\infty,1} + H^h_{\infty,2}$$
$$+ \frac{h^2}{24}(\{H^h_{\infty,2}, H^h_{\infty,2}, H^h_{\infty,1}\} - \{H^h_{\infty,1}, H^h_{\infty,1}, H^h_{\infty,2}\}) + \cdots$$

so that the combined method is of order 2 if $\psi^{[1]}$ and $\psi^{[2]}$ are of order 2. Furthermore note that if $\psi^{[1]}$, $\psi^{[2]}$ are symmetric methods so is ψ. This is a trivial consequence of the rule for finding the adjoint of a composition.

This splitting has been used by Wisdom and Holman (1991) in their simulation of the orbits of the outer planets for 10^9 years. The split Hamiltonians H_i respectively correspond to the Keplerian motion of the planets and to the perturbations such as attraction between planets etc.

12.5 Extension to the non-Hamiltonian case

The techniques considered in this chapter are not necessarily confined to Hamiltonian systems. Autonomous general systems (3.1) are specified by a vector field **F**. The Poisson bracket of the Hamiltonian case should then be replaced by the Lie (also called Poisson) bracket of vector fields (Arnold (1989), Section 39). Vector fields give rise to Lie operators that can be exponentiated to yield flows. In analysing fractional-step and related methods, exponentials can once more be combined via the BCH formula.

CHAPTER 13
High-order methods

13.1 The construction of high-order symplectic methods via Lie formalism

13.1.1 Introduction

In Chapter 8 we have provided many examples of symplectic formulae. Except for the s-stage Gauss method with order $2s$, all the methods presented there have order ≤ 4. We now investigate the construction of symplectic methods of order > 4, a task where the Lie theory of the preceding chapter can be put to good use. For simplicity, we present the construction of the various methods for the particular case of autonomous Hamiltonians. Once the tableaux of the RK, PRK and RKN methods have been obtained, it is of course possible to apply them to nonautonomous Hamiltonians. Alternatively we may reformulate nonautonomous problems in autonomous form (cf. Remark 4.1).

13.1.2 Yoshida's first approach: order 4

Let us consider a symplectic, symmetric, order-2 method $\psi_{h,H}^{[B]}$, that we shall call *the basic method*. Some choices of $\psi_{h,H}^{[B]}$ are as follows.

1. The midpoint rule (3.18).

2. If H is separable, the PRK leap-frog method $(1,0)[1/2,1/2]$ in (8.9), or $[1/2,1/2](1,0)$ in (8.11). Recall that if $(1,0)[1/2,1/2]$ is used on a partitioned system with quadratic kinetic energy (6.12), then the result is the Störmer-Verlet RKN method (8.20).

3. If H has been written in the form $H = H_1 + H_2$ and $\psi_{h,H_1}^{[1]}$ and $\psi_{h,H_2}^{[2]}$ are symplectic, symmetric, order-2 methods for S_{H_1} and S_{H_2}, the fractional-step method (12.15).

Yoshida's original presentation (Yoshida (1990)) only considers the situation where the basic method is $(1,0)[1/2, 1/2]$. However the specific nature of $\psi_{h,H}^{[B]}$ plays no role in future developments and it is convenient to leave open the choice of $\psi_{h,H}^{[B]}$.

Yoshida (1990) (see also Qin and Zhu (1992)) considers the concatenation

$$\psi_{h,H}^{[4]} = \psi_{x_1 h, H}^{[B]} \psi_{x_0 h, H}^{[B]} \psi_{x_1 h, H}^{[B]} \qquad (13.1)$$

and tries to determine the weights x_0, x_1 to ensure that $\psi_{h,H}^{[4]}$ is an order-4 method. Note that (13.1) is symplectic and symmetric for all choices of x_0 and x_1. Let $H_\infty^{[B]}$ the (formal power series for the) modified Hamiltonian for the basic method, i.e. $\phi_{h,H_\infty^{[B]}} = \psi_{h,H}^{[B]}$ (as formal power series), and assume that the expansion of $H_\infty^{[B]}$ in powers of h is

$$H_\infty^{[B]} = \alpha_0 + h^2 \alpha_2 + h^4 \alpha_4 + \cdots, \qquad (13.2)$$

with α_i functions independent of h defined in phase space (the dependence on h of the modified Hamiltonians will not be reflected in the notation). By consistency $\alpha_0 = H$; only even powers of h appear in (13.2) because $\psi_{h,H}^{[B]}$ has been taken to be symmetric. Now we use (12.12) as in the analysis of (12.15) to find the modified Hamiltonian $H_\infty^{[4]}$ for (13.1)

$$H_\infty^{[4]} = (x_0 + 2x_1)\alpha_0 + h^2(x_0^3 + 2x_1^3)\alpha_1 + O(h^4);$$

(the terms of (12.12) involving commutators only contribute to the $O(h^4)$ residual because the leading commutator $[\alpha_0, \alpha_0, \alpha_0]$ vanishes by skew-symmetry). Hence (13.1) will be of order 4 if and only if

$$x_0 + 2x_1 = 1, \qquad x_0^3 + 2x_1^3 = 0;$$

a system with the unique real solution

$$x_1 = \frac{1}{3}(2 + 2^{1/3} + 2^{-1/3}), \qquad x_0 = 1 - 2x_1.$$

If the basic method is the midpoint rule as in item 1 above, then $\psi_{h,H}^{[4]}$ is the fourth-order, diagonally implicit method (8.4). If the basic method is as in item 2 above, then we recover the method (8.15). This explains the coincidence of the weights in (8.4) and (8.15).

13.1.3 Yoshida's first approach: order 2r

After having obtained a symplectic, symmetric method $\psi^{[4]}_{h,H}$ of order 4, Yoshida goes on to look for a combination

$$\psi^{[6]}_{h,H} = \psi^{[4]}_{y_1 h,H} \psi^{[4]}_{y_0 h,H} \psi^{[4]}_{y_1 h,H}$$

of order 6. By proceeding as above, it is found that the weights y_0, y_1 should satisfy

$$y_0 + 2y_1 = 1, \qquad y_0^5 + 2y_1^5 = 0$$

with solution

$$y_1 = (2 - 2^{1/5})^{-1}, \qquad y_0 = 1 - 2y_1.$$

The construction can be iterated: once a symplectic, symmetric order-$2r$ method $\psi^{[2r]}_{h,H}$ has been found

$$\psi^{[2r+2]}_{h,H} = \psi^{[2r]}_{w_1 h,H} \psi^{[2r]}_{w_0 h,H} \psi^{[2r]}_{w_1 h,H} \qquad (13.3)$$

is symplectic, symmetric and has order $2r + 2$ provided that

$$w_1 = (2 - 2^{1/(2r+1)})^{-1}, \qquad w_0 = 1 - 2w_1.$$

13.1.4 Existence of symplectic methods of arbitrarily high orders

Obviously, the order-$(2r + 2)$ method (13.3) is a concatenation of 3^r substeps of the basic method $\psi^{[B]}_{h,H}$. Choosing $\psi^{[B]}_{h,H}$ as in items 1 or 2 above we conclude:

Theorem 13.1 *There are symplectic, diagonally implicit Runge-Kutta methods, symplectic, explicit partitioned Runge-Kutta and symplectic, explicit Runge-Kutta-Nyström methods of arbitrarily high order.*

The complexity of (13.3) grows exponentially with r and hence (13.3), for $r \geq 2$ say, is only of theoretical interest. Yoshida (1990) considers a second, more practical methodology for deriving high-order symplectic methods. This is considered next.

13.1.5 Yoshida's second approach

We start again with the basic method $\psi^{[B]}_{h,H}$, but now consider at once symmetric compositions

$$\psi^{[m]}_{h,H} = \psi^{[B]}_{w_m h,H} \cdots \psi^{[B]}_{w_1 h,H} \psi^{[B]}_{w_0 h,H} \psi^{[B]}_{w_1 h,H} \cdots \psi^{[B]}_{w_m h,H}. \qquad (13.4)$$

The modified Hamiltonian $H_\infty^{[m]}$ of (13.4), $m = 1, 2, \ldots$, can be obtained, via (12.12), from the modified Hamiltonian $H_\infty^{[m-1]}$ in view of the relation

$$\psi_{h,H}^{[m]} = \psi_{w_m h, H}^{[B]} \psi_{h,H}^{[m-1]} \psi_{w_m h, H}^{[B]}.$$

The expansion in powers of h of $H_\infty^{(m)}$ contains not only the functions α_i in (13.2) but also commutators like

$$\begin{aligned} \beta_4 &= [\alpha_0, \alpha_0, \alpha_2], \\ \beta_6 &= [\alpha_0, \alpha_0, \alpha_4], \\ \gamma_6 &= [\alpha_2, \alpha_2, \alpha_0], \\ \delta_6 &= [\alpha_0, \alpha_0, \alpha_0, \alpha_0, \alpha_2]. \end{aligned}$$

Indeed, for $m = 0, 1, 2, \ldots$, we may write

$$\begin{aligned} H_\infty^{[m]} &= A_{0,m} \alpha_0 + h^2 A_{2,m} \alpha_2 + h^4 (A_{4,m} \alpha_4 + B_{4,m} \beta_4) \\ &\quad + h^6 (A_{6,m} \alpha_6 + B_{6,m} \beta_6 + C_{6,m} \gamma_6 + D_{6,m} \delta_6) \\ &\quad + O(h^8), \end{aligned}$$

where the coefficients can be explicitly found from the recurrences

$$\begin{aligned} A_{0,m+1} &= A_{0,m} + 2w_{m+1}, \\ A_{2,m+1} &= A_{2,m} + 2w_{m+1}^3, \\ A_{4,m+1} &= A_{4,m} + 2w_{m+1}^5, \\ A_{6,m+1} &= A_{6,m} + 2w_{m+1}^7, \\ B_{4,m+1} &= B_{4,m} + \frac{1}{6}(A_{0,m}^2 w_{m+1}^3 - A_{0,m} A_{2,m} w_{m+1}) \\ &\quad - \frac{1}{6}(A_{2,m} w_{m+1}^2 - A_{0,m} w_{m+1}^4), \\ B_{6,m+1} &= B_{6,m} + \frac{1}{6}(A_{0,m}^2 w_{m+1}^5 - A_{0,m} A_{4,m} w_{m+1}) \\ &\quad - \frac{1}{6}(A_{4,m} w_{m+1}^2 - A_{0,m} w_{m+1}^6), \\ C_{6,m+1} &= C_{6,m} + \frac{1}{6}(A_{2,m}^2 w_{m+1} - A_{0,m} A_{2,m} w_{m+1}^3) \\ &\quad - \frac{1}{6}(A_{0,m} w_{m+1}^6 - A_{2,m} w_{m+1}^4), \\ D_{6,m+1} &= D_{6,m} + \frac{7}{360}(A_{2,m} w_{m+1}^4 - A_{0,m} w_{m+1}^6) \\ &\quad + \frac{7}{180}(A_{0,m} A_{2,m} w_{m+1}^3 - A_{0,m}^2 w_{m+1}^5) \end{aligned}$$

$$-\frac{1}{45}(A_{0,m}^3 w_{m+1}^4 - A_{0,m}^2 A_{2,m} w_{m+1}^2)$$
$$-\frac{1}{360}(A_{0,m}^4 w_{m+1}^3 - A_{0,m}^3 A_{2,m} w_{m+1})$$
$$-\frac{1}{6}(B_{4,m} w_{m+1}^2 + A_{0,m} B_{4,m} w_{m+1}),$$

valid for $m = 0, 1, 2, \ldots$, with initial conditions
$$A_{k,0} = w_0^{k+1}, \quad k = 0, 2, 4, 6,$$
$$B_{4,0} = B_{6,0} = C_{6,0} = D_{6,0} = 0.$$

For $\psi_{h,H}^{[m]}$ to be an integrator of order 6 we have four conditions
$$A_{0,m} = 1, \quad A_{2,m} = A_{4,m} = B_{4,m} = 0,$$
and for order 8 we also impose
$$A_{6,m} = B_{6,m} = C_{6,m} = D_{6,m} = 0.$$

Yoshida solves these order equations numerically to find methods of order 6 with $m = 3$ (four order equations for four unknown weights w_i). He then obtains in a similar way methods of order 8 with $m = 7$ (eight order conditions for eight unknown weights). A method of order 8 is given by

$$\begin{aligned}
w_1 &= 0.102799849391985 E0, \\
w_2 &= -0.196061023297549 E1, \\
w_3 &= 0.193813913762276 E1, \\
w_4 &= -0.158240635368243 E0, \\
w_5 &= -0.144485223686048 E1, \\
w_6 &= 0.253693336566229 E0, \\
w_7 &= 0.914844246229740 E0;
\end{aligned} \qquad (13.5)$$

the value w_0 is recovered from the consistency equation
$$A_0 = w_0 + 2 \sum_{i=1}^{m} w_i = 1.$$

As mentioned before, Yoshida's original paper assumes that the basic method $\psi_{h,H}^{[B]}$ is the leap-frog PRK $(1,0)[1/2, 1/2]$. In this case (13.4) is the PRK method (see (8.16))

$$(w_m, \ldots, w_1, w_0, w_1, \ldots, w_m, 0)$$
$$\left[\frac{w_m}{2}, \ldots, \frac{w_1 + w_0}{2}, \frac{w_1 + w_0}{2}, \ldots, \frac{w_m}{2}\right]. \qquad (13.6)$$

With the coefficients (13.5), the tableaux (13.6) specify an order 8, symplectic PRK method with 16 stages. Due to the FSAL property, per step, 15 evaluations of f and g are required.

Other useful related references are Suzuki (1991) and Forest (1992).

13.2 High-order Runge-Kutta-Nyström methods

13.2.1 Order 7 methods

We now examine the construction of RK, RKN or PRK methods via traditional order conditions (Chapter 4), rather than via Lie formalism. Our aim is to illustrate the underlying difficulties: we see from Tables 7.1 and 7.2 that the number of order conditions for order r increases dramatically with r.

Calvo and Sanz-Serna (1993c) consider the task of constructing an order-7, symplectic, explicit RKN method. From Table 7.2 we see that there are 25 order conditions to be imposed. Since the tableau (8.18) contains $2s$ free parameters, it seems that 13 or more stages are required. A technique for reducing the number of stages is the use of *simplifying assumptions*. A simplifying assumption is a relation on the tableau coefficients that implies that some of the standard order conditions become a consequence of other order conditions. For instance Lemma 7.4 shows that the symplecticness conditions (6.15)–(6.16) are simplifying assumptions. We also pointed out in Remark 4.3 the role of (4.12) as a simplifying assumption.

A familiar simplifying assumption for RKN methods is given by

$$\sum_{j=1}^{s} \alpha_{ij} = \frac{\gamma_i^2}{2}, \qquad i = 1, \ldots, s, \qquad (13.7)$$

(Hairer et al. (1987), Chapter II, Lemma 13.14). When (13.7) holds it is possible to disregard the order conditions associated with special Nyström rooted trees with two or more vertices where at least one end vertex is fat. It is easy to see that when both (13.7) and the symplecticness conditions (6.15)–(6.16) hold, then for order 7 only 12 order conditions are necessary. For instance, for special Nyström rooted trees with four vertices (see Fig. 4.4), the order condition of $\sigma\nu\rho\tau_{4,1}$ is equivalent to the order condition for $\sigma\nu\rho\tau_{4,2}$ by symplecticness, and the latter order condition can be disregarded by (13.7).

Calvo and Sanz-Serna notice that if in the explicit, symplectic

RKN method (8.18) with $s \geq 2$ one sets

$$\gamma_1 = 0, \qquad \gamma_s = 1, \qquad (13.8)$$

and

$$\begin{aligned} b_1 &= \gamma_2/2, \\ b_i &= (\gamma_{i+1} - \gamma_{i-1})/2, \qquad i = 2, \ldots, s-1, \\ b_s &= (1 - \gamma_{s-1})/2, \end{aligned} \qquad (13.9)$$

then (13.7) is satisfied. Note that (13.8) and (13.9) only leave $s-2$ free parameters $\gamma_2, \ldots, \gamma_{s-1}$ in (8.18). However the FSAL property holds and only $s-1$ evaluations per step are required. Furthermore the order is ≥ 2 for any choice of the free parameters because the order conditions for order ≥ 2, i.e. $\sum b_i = 1$, $\sum b_i \gamma_i = 1/2$ are implied by the structure of the tableau (8.18) and the relations (13.8)–(13.9).

As metioned above, order 7 demands 12 order conditions, and two of them are already satisfied. Therefore $\gamma_2, \ldots, \gamma_{s-1}$ should satisfy 10 conditions. This suggests $s \geq 12$. The choice $s = 12$ leaves no freedom to 'tune' the formula and Calvo and Sanz-Serna take $s = 13$. This leads to a one-parameter family of explicit, symplectic methods $\psi^{[7]}$ of order 7, with 13 stages and 12 function evaluations per step.

13.2.2 Order 8 out of order 7

As in Subsection 8.4.5, the composition of $\psi^{[7]}$ with its adjoint $\bar{\psi}^{[7]}$ results in a symmetric, symplectic, order 8 method

$$\psi^{[8]}_{h,H} = \psi^{[7]}_{h/2,H} \bar{\psi}^{[7]}_{h/2,H}. \qquad (13.10)$$

The overall method has 26 stages and (due to the FSAL property) 24 function evaluations per step. However, for reasons detailed in Subsection 8.4.5, the method yields order 8 output at the midway point $t_n + h/2$, so that order 8 output is available after every 12 function evaluations.

Calvo and Sanz-Serna determine the free parameter in $\psi^{[7]}_{h/2,H}$ to minimize the RKN error constants of $\psi^{[8]}_{h,H}$ as in Subsection 8.5.3. This leads to the following abscissae for $\psi^{[7]}$:

$$\begin{aligned} \gamma_2 &= 0.60715821186110352503, \\ \gamma_3 &= 0.96907291059136392378, \\ \gamma_4 &= -0.10958316365513620399, \end{aligned}$$

$$\gamma_5 = 0.05604981994113413605,$$
$$\gamma_6 = 1.30886529918631234010,$$
$$\gamma_7 = -0.11642101198009154794, \quad (13.11)$$
$$\gamma_8 = -0.29931245499473964831,$$
$$\gamma_9 = -0.16586962790248628655,$$
$$\gamma_{10} = 1.22007054181677755238,$$
$$\gamma_{11} = 0.20549254689579093228,$$
$$\gamma_{12} = 0.86890893813102759275.$$

13.2.3 Connection with the Lie formalism

The method (8.18), (13.8), (13.9), is in fact a composition

$$\psi_{h,H} = \psi^{[2]}_{(\gamma_s-\gamma_{s-1})h,H} \cdots \psi^{[2]}_{(\gamma_2-\gamma_1)h,H}$$

based on the RKN method $\psi^{[2]}_{h,H}$ with tableau

0	0	0
1	1/2	0
	1/2	1/2
	1/2	0

This, in turn coincides with the RKN scheme induced by the PRK leap-frog method $[1/2, 1/2](1, 0)$ in (8.11). When composing with the adjoint, we find ($\psi^{[2]}_{h,H}$ is symmetric):

$$\bar\psi_{h/2,H}\psi_{h/2,H} = \psi^{[2]}_{(\gamma_2-\gamma_1)h/2,H} \cdots \psi^{[2]}_{(\gamma_{s-1}-\gamma_{s-2})h/2,H}$$
$$\times \left[\psi^{[2]}_{(\gamma_s-\gamma_{s-1})h/2,H}\psi^{[2]}_{(\gamma_s-\gamma_{s-1})h/2,H}\right]$$
$$\times \psi^{[2]}_{(\gamma_{s-1}-\gamma_{s-2})h/2,H} \cdots \psi^{[2]}_{(\gamma_2-\gamma_1)h/2,H}.$$

One can use the BCH formula (12.11) to find the modified Hamiltonian for the term in square brackets. In doing this all the terms involving commutators vanish. Once this modified Hamiltonian has been found we may iteratively apply (12.12) to find the modified Hamiltonian for $\bar\psi_{h,H}\psi_{h,H}$. Hence the method derived here via simplifying assumptions may alternatively be constructed via Lie formalism.

13.3 A comparison of order 8 symplectic integrators

13.3.1 Methods being compared

Some numerical experiments are now presented. We have implemented the following order 8 RKN methods:

Calvo8 The explicit, symplectic RKN method with 26 stages, 24 function evaluations per step discussed in the preceding section. (More precisely the method is given by (13.10), with $\psi^{[7]}$ defined by (8.18), (13.8)–(13.9), (13.11).) The formula has optimized error constants and has been implemented with constant step sizes.

Dormand8 The explicit, *nonsymplectic* embedded pair with nine stages in Table 1 of Dormand et al. (1987b). The algorithm requires eight function evaluations per step and has been implemented (of course with variable step sizes) as described in Chapter 5. The embedded method used for error estimation has order 6. This algorithm is included in the comparisons as a good example of state-of-the-art, optimized, standard integrator.

Gauss8 The *implicit,* symplectic RKN method induced by the four-stage Gauss method (see Subsection 3.3.2). This is implemented with constant step sizes, with the algebraic equations solved by functional iteration as described in Chapter 5. The algorithm requires four function evaluations per inner iteration; the number of evaluations per step then depends on the number of inner iterations required to solve the algebraic equations.

Yoshida8 The RKN method induced by the explicit, symplectic PRK method of order 8 derived by Yoshida and described earlier in this chapter (formulae (13.5)–(13.6)); 15 function evaluations per step are required.

Remark 13.1 Yoshida (1990) provides alternative choices for the weights in (13.5). Amongst Yoshida's methods, (13.5) is that for

Table 13.1. *Eighth-order methods being compared*

Method	N	A	$\dot{A}$	$A \cdot N^9$	$\dot{A} \cdot N^9$
Calvo8	24	$1. \times 10^{-5}$	$1. \times 10^{-5}$	$3. \times 10^7$	$3. \times 10^7$
Dormand8	8	$8. \times 10^{-7}$	$8. \times 10^{-7}$	$1. \times 10^2$	$1. \times 10^2$
Gauss8	–	$2. \times 10^{-6}$	$2. \times 10^{-6}$	–	–
Yoshida8	15	$2. \times 10^{-3}$	$4. \times 10^{-3}$	$9. \times 10^7$	$1. \times 10^8$

which the induced RKN method has smallest error constants. □

In Table 13.1 we have listed, for each method, the number N of evaluations per step, the error constants A and $\dot{A}$ and scaled error constants $A \cdot N^9$, $\dot{A} \cdot N^9$. We note that both explicit symplectic methods have extremely large scaled error constants. Furthermore, the method Calvo8, where some optimization was performed, has better scaled error constants than Yoshida8, a method in whose derivation no free parameters were allowed: the number of stages is the bare minimum to satisfy the necessary order conditions.

13.3.2 Results: Kepler's problem

The methods presented above were tested in Kepler's problem as in Subsection 9.1.2. In Fig. 13.1 we show error against CPU time in seconds on a SUN Sparc IPX workstation. The most efficient method is the implicit Gauss8 (dotted line and × signs, $h = 2\pi/128$, $2\pi/256$). This is followed by Calvo8 (solid line and stars, $h = 2\pi/32$, $2\pi/64$, $2\pi/128$) and the nonsymplectic Dormand8 (dashed line and + signs, tolerances 10^{-9} to 10^{-13}). The method Yoshida8 (dash-dot line, circles, $h = 2\pi/128$, $2\pi/256$) is the least efficient.

As it was the case with order-4 methods (Subsection 9.1.2), the relative efficiency of order-8 *symplectic* algorithms on this test problem is easily explained in terms of error constants. Furthermore, when comparing the three symplectic algorithms on the one hand with the nonsymplectic method on the other, it is clear that, once again, the symplectic algorithms' performance cannot be explained in terms of the size of the local error (see Subsection 9.1.2). However, now the advantages of symplectic formulae are not so marked as they were in the order-4 case. The order-8 symplectic algorithms tested are either implicit or require many stages (leading to large scaled error constants) and these drawbacks offset the advantages of better error propagation deriving from symplecticness.

It is useful to compare Fig. 13.1 with Fig. 9.1: the *worst* order-8 performance (Yoshida8) improves on the *best* order-4 method (Calvo). For the situation at hand (smooth problem and high accuracy) high order certainly pays.

Fig. 13.2 corresponds to the runs in Fig. 13.1, but now work is measured in number of function evaluations. The nonsymplectic Dormand8 turns out to be the most efficient method. The conclusion is that for expensive problems requiring highly accurate quan-

A COMPARISON OF ORDER 8 SYMPLECTIC INTEGRATORS 175

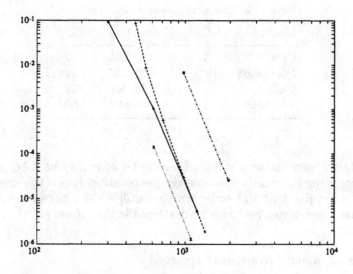

Figure 13.1. *Error as a function of CPU time after 21870 periods*

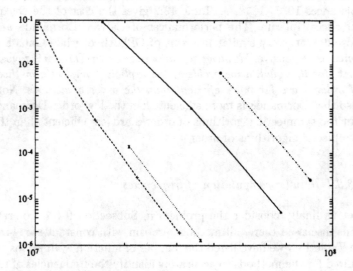

Figure 13.2. *Error as a function of number of evaluations after 21870 periods*

Table 13.2. *Cost of the cheapest successful run*

Method	h/Tol	Evaluations	CPU
Calvo8	1	4800001	60
Dormand8	10^{-5}	2038737	64
Gauss8	1	7485665	163
Yoshida8	1	3000001	60

titative information a standard high-order code may be a better choice than currently available symplectic algorithms. Of course, it is possible that high-order symplectic algorithms more efficient than those tested here may be developed in the future.

13.3.3 Results: Hénon-Heiles problem

We now apply the order-8 methods to the Hénon-Heiles problem considered in Subsection 9.1.3. We run the constant step-size algorithms with $h = 2, 1, 1/2, \ldots$ and stop halving h when the correct solution is produced. For the variable step-size Dormand8 we use tolerances $10^{-4}, 10^{-5}, \ldots$ Table 13.2 gives the cost of the cheapest successful run. The performances of Calvo8, Dormand8 and Yoshida8 are very similar in terms of CPU time, while Gauss8 is twice as expensive. However a comparison with Table 9.2 reveals that, *for this qualitative problem, the explicit, symplectic methods of order 4 are far more efficient than the order 8 methods*. Note also that Dormand8 is more efficient than the low-order Dormand, but the symplectic algorithms of order 8 are less efficient than the symplectic algorithms of order 4.

13.3.4 Results: computation of frequencies

Let us finally consider the problem in Subsection 9.1.4. For reasons discussed there, all methods are run with constant step size. In Table 13.3 we have listed the number of function evaluations required for the methods to accurately identify the frequencies of the problem. Data for the order-4 Calvo and Dormand are also given. The symplectic order-8 methods are more efficient than Dormand and Dormand8. Clearly the most efficient method is Calvo.

Table 13.3. *Evaluations in the cheapest accurate run*

Method	h	Evaluations
Calvo	1/8	262113
Dormand	1/64	1572673
Calvo8	1/4	786337
Dormand8	1/32	2096897
Gauss8	1/4	737352
Yoshida8	1/8	982920

13.3.5 Conclusions

The experiments we have reported suggest that symplectic order-4 methods clearly improve on conventional order-4 or order-8 algorithms in problems where *the qualitative behaviour of the solution is of interest*.

For problems where *high accuracy* is required high-order methods should be used. Current high-order symplectic methods are either implicit or possess too many stages relative to the size of their error constants. Accordingly, in problems with expensive function evaluations, currently available symplectic methods may not be competitive with the best high-order nonsymplectic algorithms.

CHAPTER 14

Extensions

14.1 Partitioned Runge-Kutta methods for nonseparable Hamiltonian systems

The last chapter is devoted to some extensions of the preceding material. Some of the developments that follow are very recent and a thorough treatment is not yet possible.

So far in this book the application of the PRK method (3.24) has been restricted to partitioned (not necessarily Hamiltonian) systems of the special form (3.23). It is however possible to integrate any partitioned system (3.22) by the PRK method defined by the tableaux (3.24). The relations (3.25)–(3.26) should be replaced by

$$\mathbf{P}_i = \mathbf{p}^n + h_{n+1} \sum_{j=1}^{s} a_{ij} \mathbf{f}(\mathbf{P}_j, \mathbf{Q}_j, t_n + C_j h_{n+1}),$$

$$\mathbf{Q}_i = \mathbf{q}^n + h_{n+1} \sum_{j=1}^{s} A_{ij} \mathbf{g}(\mathbf{P}_j, \mathbf{Q}_j, t_n + C_j h_{n+1}),$$

and the equations defining $\mathbf{p}^{n+1}$ and $\mathbf{q}^{n+1}$ should also be modified accordingly.

With these modifications, (3.24) can be applied to the integration of general Hamiltonian systems, i.e. to Hamiltonian systems where H is not separable (see (6.8)). Suris (1990) and Abia and Sanz-Serna (1993) point out that the method is then symplectic if, in addition to (6.9),

$$b_i = B_i, \qquad i = 1, \ldots, s.$$

This follows easily from (6.10) and (6.7).

Example 14.1 Sun (1992b) (see also Jay (1993)) has proved that the combination of the Lobatto IIIA and Lobatto IIIB tableaux (Hairer and Wanner (1991), Chapter IV.5) gives rise to a PRK method that is symplectic for general Hamiltonians H. □

14.2 Canonical B-series

With the notation of Section 4.2, a *B-series* (Hairer and Wanner (1974), Hairer *et al.* (1987) Chapter II.11) is a (formal) series

$$\mathbf{y} + \sum_{m=1}^{\infty} \frac{h^m}{m!} \sum_{\rho\tau \in T_m} c(\rho\tau)\mathcal{F}(\rho\tau)(\mathbf{y}) \qquad (14.1)$$

where, for each rooted tree $\rho\tau$, $c(\rho\tau)$ is a real number. According to (4.8), the expansion in powers of h of the flow $\phi_{h,\mathbf{F}}$ is a B-series; namely the B-series with $c(\rho\tau) = \alpha(\rho\tau)$ for each rooted tree $\rho\tau$. In a similar manner, the expansion of an RK solution (4.7) is the B-series given by

$$c(\rho\tau) = \alpha(\rho\tau)\gamma(\rho\tau)\Phi(\rho\tau). \qquad (14.2)$$

Other numerical integrators such as multiderivative RK methods (Hairer *et al.* (1987) Chapter II.12) also give rise to B-series. For reasons presented in Hairer *et al.* (1987) Chapter II.11, the investigation of the properties of a method via the corresponding B-series is often advantageous.

It is then of clear interest to ascertain under which condition (14.1) is symplectic for each Hamiltonian system. Calvo and Sanz-Serna have very recently shown (1993d) that (14.1) is symplectic if and only if, in the situation in Fig. 7.1,

$$\frac{c(\rho\tau_i)}{\alpha(\rho\tau_i)\gamma(\rho\tau_i)} + \frac{c(\rho\tau_j)}{\alpha(\rho\tau_j)\gamma(\rho\tau_j)}$$
$$= \frac{c(\rho\tau_I)}{\alpha(\rho\tau_I)\gamma(\rho\tau_I)} \frac{c(\rho\tau_J)}{\alpha(\rho\tau_J)\gamma(\rho\tau_J)} \qquad (14.3)$$

for each pair of rooted trees $\rho\tau_I$, $\rho\tau_J$.

Example 14.2 As noted above, for the true flow, $c(\rho\tau_i) = \alpha(\rho\tau_i)$. Then the condition (14.3) becomes (7.1), a formula whose validity was proved in Lemma 7.1. This shows, once more and at least formally, the symplecticness of Hamiltonian flows. □

Example 14.3 If (14.1) comes from an RK method, then (14.2) holds. In this case (14.3) becomes (7.2), a relation that holds for symplectic RK methods as shown in Lemma 7.1. □

The characterization of symplectic B-series can be used to discover whether a given method is symplectic or otherwise. For instance Hairer *et al.* (1993) use the result in Calvo and Sanz-Serna (1993d) to show that there is no symplectic, multiderivative RK method.

14.3 Conjugate symplectic methods. The trapezoidal rule

The RK method with tableau

$$\begin{array}{c|cc} 0 & 0 & \\ 1/2 & 1/2 & 1/2 \\ \hline & 1/2 & 1/2 \end{array}$$

is the well-known trapezoidal rule

$$\mathbf{y}^{n+1} = \mathbf{y}^n + \frac{h}{2}\left[\mathbf{F}(\mathbf{y}^n, t_n) + \mathbf{F}(\mathbf{y}^{n+1}, t_{n+1})\right]. \qquad (14.4)$$

This is sometimes confused with the midpoint rule (3.18)–(3.19). While both methods coincide for linear, constant-coefficient problems ($\mathbf{F}(\mathbf{y}, t) = A\mathbf{y}$, A a constant matrix), they are in general different. For instance the midpoint rule is symplectic while the trapezoidal rule does not satisfy the symplecticness requirement (6.5).

There is a close relation between numerical results of the midpoint and trapezoidal rules when applied with constant step sizes to autonomous problems (Dahlquist (1975)). Assume that $\mathbf{y}_M^0, \mathbf{y}_M^1, \mathbf{y}_M^2 \ldots$ is a midpoint solution so that

$$\mathbf{y}_M^{n+1} - \mathbf{y}_M^n = h\mathbf{F}\left(\frac{1}{2}(\mathbf{y}_M^n + \mathbf{y}_M^{n+1})\right). \qquad (14.5)$$

We also write this equation with n replaced by $n+1$

$$\mathbf{y}_M^{n+2} - \mathbf{y}_M^{n+1} = h\mathbf{F}\left(\frac{1}{2}(\mathbf{y}_M^{n+1} + \mathbf{y}_M^{n+2})\right). \qquad (14.6)$$

On adding (14.5) and (14.6) and rearranging, we obtain

$$\left[\frac{1}{2}(\mathbf{y}_M^{n+1} + \mathbf{y}_M^{n+2})\right] - \left[\frac{1}{2}(\mathbf{y}_M^n + \mathbf{y}_M^{n+1})\right]$$
$$= \frac{h}{2}\left(\mathbf{F}\left(\frac{1}{2}(\mathbf{y}_M^n + \mathbf{y}_M^{n+1})\right) + \mathbf{F}\left(\frac{1}{2}(\mathbf{y}_M^{n+1} + \mathbf{y}_M^{n+2})\right)\right),$$

i.e., the averages

$$\mathbf{y}_T^{n+1} := \frac{1}{2}(\mathbf{y}_M^n + \mathbf{y}_M^{n+1})$$

satisfy the trapezoidal rule equation (14.4). On the other hand

$$\mathbf{y}_M^{n+1} = 2\mathbf{y}_T^{n+1} - \mathbf{y}_M^n = 2\mathbf{y}_T^{n+1} - (\mathbf{y}_M^{n+1} - h\mathbf{F}(\mathbf{y}_T^{n+1})),$$

and hence
$$\mathbf{y}_M^{n+1} = \mathbf{y}_T^{n+1} + \frac{h}{2}\mathbf{F}(\mathbf{y}_T^{n+1}).$$
The conclusion is that the mapping
$$\chi(\mathbf{y}) = \mathbf{y} + \frac{h}{2}\mathbf{F}(\mathbf{y})$$
(that depends on $\mathbf{F}$ and h) maps trapezoidal rule solutions $\mathbf{y}_T^0$, $\mathbf{y}_T^1$, ... into midpoint rule solutions $\mathbf{y}_M^0$, $\mathbf{y}_M^1$, ... For h small χ is invertible so that, conversely, midpoint rule solutions are taken by χ^{-1} into trapezoidal rule solutions. In symbols
$$\begin{aligned}\psi^{[T]} &= \chi^{-1}\psi^{[M]}\chi,\\ \psi^{[M]} &= \chi\psi^{[T]}\chi^{-1}.\end{aligned}$$

Thus, while the trapezoidal rule is not symplectic it is the result of changing variables in a symplectic method. Methods that result from changing variables in a symplectic scheme are called, following Stofer (1988a), (1988b), *conjugate canonical or conjugate symplectic*. They obviously share many of the properties of symplectic methods.

14.4 Constrained systems

The pendulum provides a good example of a constrained system. The description of the pendulum motion in Chapter 1 uses as dependent variables the angle q between the rod and the downward vertical axis and the corresponding angular velocity p. As an alternative, it is possible to describe the pendulum in terms of the cartesian coordinates x, y of the blob and corresponding cartesian velocities v_x, v_y. The equations of motion are then (we assume that the mass of the blob, the length of the rod and the acceleration of gravity are all unity)

$$\dot{v}_x = -x\lambda, \tag{14.7}$$
$$\dot{v}_y = -1 - y\lambda, \tag{14.8}$$
$$\dot{x} = v_x, \tag{14.9}$$
$$\dot{y} = v_y. \tag{14.10}$$

Here λ is a (time-dependent) scalar and $(-x\lambda, -y\lambda)$ represents the force, parallel to the rod, that the rod exerts on the blob. There are four equations for five unknown functions v_x, v_y, x, y, λ and a further relation is required. This obviously is given by the

constraint
$$x^2 + y^2 - 1 = 0. \tag{14.11}$$

On differentiating this coordinate constraint with respect to t, we obtain a constraint for the velocities
$$xv_x + yv_y = 0 \tag{14.12}$$

(the velocity vector (v_x, v_y) is orthogonal to the rod). A further differentiation yields
$$v_x^2 + v_y^2 + x(-x\lambda) + y(-1 - y\lambda) = 0$$
or
$$\lambda = v_x^2 + v_y^2 - y \tag{14.13}$$

(the tension in the rod balances the centrifugal and gravitational forces).

In the four-dimensional space of the variables (v_x, v_y, x, y), the constraints (14.11), (14.12) define a two-dimensional manifold $\mathcal{M}$. Given initial values on this manifold $(v_x(0), v_y(0), x(0), y(0)) \in \mathcal{M}$, the equations (14.7)–(14.10) (with λ given by (14.13)) can be used to evolve on $\mathcal{M}$ the point (v_x, v_y, x, y) to a final position $(v_x(t), v_y(t), x(t), y(t))$. In this evolution (14.11) and (14.12) are conserved quantities. In this way, a flow ϕ_t on $\mathcal{M}$ is defined. Furthermore ϕ_t preserves the differential 2-form
$$dv_x \wedge dx + dv_y \wedge dy. \tag{14.14}$$

This can be understood by considering the pendulum as a limit case of the situation where the rod is attached to a stiffer and stiffer spring of unit length. Each spring system, being Hamiltonian in $\mathcal{R}^4$, preserves (14.14) and preservation then holds for the limit system.

Numerically, the system (14.7)–(14.10) supplemented by *either* (14.11) *or* (14.12) *or* (14.13) (so as to have five equations for five unknowns) can be integrated by a *differential-algebraic integrator*
$$(v_x^n, v_y^n, x^n, y^n, \lambda^n) \to (v_x^{n+1}, v_y^{n+1}, x^{n+1}, y^{n+1}, \lambda^{n+1}),$$

see, e.g., Brenan *et al.* (1989), Hairer *et al.* (1989), Hairer and Wanner (1991), Chapter VI.5. The question arises whether this evolution also preserves (14.14). Recent results in this direction for general mechanical systems can be seen in Leimkuhler and Reich (1992), Jay (1993), Leimkuhler and Skeel (1993), Reich (1993).

What is the relation with our treatment of the pendulum in terms of the p, q variables of Chapter 1? These variables provide a local chart (parametrization) of the manifold $\mathcal{M}$. If

$$
\begin{aligned}
x &= \sin q, \\
y &= -\cos q, \\
v_x &= p \cos q, \\
v_y &= p \sin q,
\end{aligned}
$$

then (14.11) and (14.12) are automatically satisfied and (14.7)–(14.10), (14.13) collapse into $\dot{p} = -\sin q$, $\dot{q} = p$. Furthermore, on $\mathcal{M}$

$$
\begin{aligned}
dv_x \wedge dx + dv_y \wedge dy &= (\cos q \, dp - p \sin q \, dq) \wedge (\cos q \, dq) \\
&\quad + (\sin q \, dp + p \cos q \, dq) \wedge (\sin q \, dq) \\
&= (\cos^2 q + \sin^2 q)(dp \wedge dq) = dp \wedge dq,
\end{aligned}
$$

so that symplectic methods for the (p,q) formulation, when conserving $dp \wedge dq$, are also conserving (14.14).

Working with a parametrization (such as (p,q)) reduces the dimensionality of the problem and avoids the need for differential-algebraic integrators. The fly in the ointment is that parametrizations are not always easy to find.

14.5 General Poisson structures

The *standard* Hamiltonian format (1.1) we have been considering throughout the book is not the most general considered in the literature. In Chapter 12 we saw that the formalism associated with the standard format can be presented in terms of the associated Poisson bracket (12.1). Switching the emphasis from the equations of motion (1.1) to the Poisson bracket provides a way for generalizing the standard format.

If $\mathcal{M}$ is a manifold, a Poisson bracket (Marsden (1992), Olver (1986)) in $\mathcal{M}$ is a mapping that associates with each pair (F, G) of smooth real-valued functions on $\mathcal{M}$ a number $\{F, G\}$ so that the properties of bilinearity, skew-symmetry, Jacobi's identity and Leibniz's rule are satisfied as in Section 12.1. With this terminology we see that in fact (12.1) defines a Poisson bracket (the standard Poisson bracket) in $\mathcal{M} = \mathcal{R}^{2d}$ (or in any subdomain Ω of $\mathcal{R}^{2d}$).

If $\mathcal{M}$ is finite-dimensional and $z_1, \ldots, z_D$ are local coordinates on $\mathcal{M}$, then each Poisson bracket $\{\cdot, \cdot\}$ has an expression

$$\{F, G\} = \sum_{i,j=1}^{D} \frac{\partial F}{\partial z_i} B_{ij}(\mathbf{z}) \frac{\partial G}{\partial z_j}, \qquad (14.15)$$

where, for each $\mathbf{z}$, $B_{ij}(\mathbf{z})$ is a $D \times D$ skew-symmetric matrix, the *structure matrix*. Furthermore, the Jacobi condition reads, in terms of the structure matrix,

$$\sum_{l,k=1}^{D} \left(B_{li} \frac{\partial B_{jk}}{\partial z_l} + B_{lj} \frac{\partial B_{ki}}{\partial z_l} + B_{lk} \frac{\partial B_{ij}}{\partial z_l} \right) = 0.$$

The standard case has $B(\mathbf{z}) = J^{-1}$, and the last relation is trivially satisfied. If $B(\mathbf{z})$, while not coinciding with the standard J^{-1}, is constant and invertible (this implies that D is even $= 2d$), then the situation is essentially the same as the standard. Hence for an interesting generalization, B should be singular and/or effectively depend on $\mathbf{z}$.

A *Poisson mapping* $\psi : \mathcal{M} \to \mathcal{M}$ is a mapping that preserves the Poisson bracket, in the sense of Remark 12.1.

In local coordinates $\mathbf{z}^* = \psi(\mathbf{z})$ is Poisson if

$$\left(\frac{\partial \psi}{\partial \mathbf{z}} \right)^T B \left(\frac{\partial \psi}{\partial \mathbf{z}} \right) = B.$$

If H is a function on $\mathcal{M}$, then we define the *Hamiltonian system (equations of motion)* associated with H as

$$\dot{F} = \{F, H\}$$

for all smooth F on $\mathcal{M}$. According to (12.5) this generalizes the situation for the standard case. When working in local coordinates z_i the flow is retrieved from the evolution of the coordinate functions (cf. (12.3))

$$\dot{z}_i = \{z_i, H\},$$

i.e.

$$\dot{\mathbf{z}} = B(\mathbf{z})\nabla H,$$

with $\nabla = (\partial/\partial z_1, \ldots, \partial/\partial z_D)$.

The flow of Hamilton's equations is a Poisson map and one looks for numerical methods that, in a like manner, are Poisson maps. For references to such *Poisson integrators* see Scovel (1991), Marsden (1992).

14.6 Multistep methods

Assume now that a D-dimensional autonomous system $\dot{\mathbf{y}} = \mathbf{F}(\mathbf{y})$ is integrated by an irreducible linear multistep method in its one-leg

version (Hairer and Wanner (1991), Chapter V.6)
$$\alpha_k \mathbf{y}^{n+k} + \cdots + \alpha_0 \mathbf{y}^n = h\mathbf{F}(\beta_k \mathbf{y}^{n+k} + \cdots + \beta_0 \mathbf{y}^n).$$
Then the method induces a mapping ψ in $\mathcal{R}^{kD}$ of the form
$$\begin{align} \mathbf{Y}^{n+1} &= \psi(\mathbf{Y}^n) \\ \mathbf{Y}^n &= (\mathbf{y}^{n+k-1}, \ldots, \mathbf{y}^n). \end{align} \tag{14.16}$$

If $D = 2d$ and the system being integrated is an autonomous standard Hamiltonian system (1.1), it is natural to ask whether (14.16) defines a Poisson mapping in $\mathcal{R}^{k2d}$ with respect to a constant structure matrix B of the form $\Lambda \otimes J^{-1}$, where Λ is $k \times k$ and only depends on the method coefficients, J is the matrix in (1.4) and $\otimes$ is the Kronecker product (see (5.5)).

Eirola and Sanz-Serna (1992) show that this is the case if and only if the method is *symmetric*, i.e.
$$\alpha_j = \alpha_{k-j}, \qquad \beta_j = \beta_{k-j}, \qquad j = 0, 1, \ldots, k.$$
Symmetric methods are used by Quinlan and Tremaine (1990) for long-time simulations of planetary orbits.

In particular, the explicit midpoint rule
$$\mathbf{y}^{n+2} - \mathbf{y}^n = 2h\mathbf{F}(\mathbf{y}^{n+1})$$
is symplectic in this extended sense. Furthermore, it can be shown (Sanz-Serna and Vadillo (1987)) that, in this particular case, the mapping $(\mathbf{y}^n, \mathbf{y}^{n+1}) \to (\mathbf{y}^{n+1}, \mathbf{y}^{n+2})$ is actually an approximation of the flow in $\mathcal{R}^{4d}$ of a standard Hamiltonian system.

14.7 Partial differential equations

The manifold $\mathcal{M}$ in Section 14.5 may also be infinite-dimensional leading to partial differential equations (PDEs). Typically, $\mathcal{M}$ consists of smooth functions $\mathbf{u}$ of a spatial variable $\mathbf{x}$. The real-valued functions on $\mathcal{M}$ are functionals $F, G, \ldots$ depending on $\mathbf{u}$ and the Poisson bracket can be written as (cf. (14.15))
$$\{F, G\} = \int \delta F B(\mathbf{u}) \, \delta G,$$
where δ denotes variational derivative, B is skew-symmetric and has to satisfy the Jacobi property. The equations of motion for the Hamiltonian functional H are
$$\frac{\partial \mathbf{u}}{\partial t} = B(\mathbf{u}) \delta H.$$

Example 14.4 We first consider (McLachlan (1992)) *nonlinear wave equations* in one space dimension. For these, $\mathbf{u} = (p(x), q(x))$ (p and q satisfy suitable boundary conditions),

$$B = \begin{bmatrix} 0 & -1 \\ 1 & 0 \end{bmatrix}$$

and

$$H(\mathbf{u}) = \int \left[\frac{1}{2}p^2 + \frac{1}{2}q_x^2 + V(q) \right] dx.$$

The equations of motion are

$$\frac{\partial p}{\partial t} = -\frac{\delta H}{\delta q} = q_{xx} - V'(q),$$
$$\frac{\partial q}{\partial t} = \frac{\delta H}{\delta p} = p,$$

or, eliminating p,

$$\frac{\partial^2 q}{\partial t^2} = \frac{\partial^2 q}{\partial x^2} - V'(q). \quad \square$$

Example 14.5 The choice $B = \partial/\partial x$,

$$H(\mathbf{u}) = \int \left(-u^3 + \frac{1}{2}u_x^2 \right) dx$$

leads to the Korteweg-de Vries equation

$$\frac{\partial u}{\partial t} = \frac{\partial}{\partial x}(-3u^2 - u_{xx}) = -6uu_x - u_{xxx}. \quad \square$$

The numerical integration of time dependent PDEs may often be conceived as consisting of two parts (see e.g. Sanz-Serna and Verwer (1989)). First the spatial derivatives are discretized by finite differences, finite elements, spectral methods, etc. to obtain a system of ordinary differential equations (ODEs) with t as the independent variable. Then this system of ODEs is integrated numerically. If the PDEs are of Hamiltonian type, one may insist that both stages preserve the Hamiltonian structure. Thus the space discretization should be carried out in such a way that the resulting system of ODEs is Hamiltonian (for a suitable Poisson bracket) and the time integration should also be carried out by a symplectic or Poisson integrator. Some references are Li and Qin (1988), Frutos *et al.* (1990), Qin and Zhang (1990), Huang (1991), Wang (1991), McLachlan (1992), Frutos and Sanz-Serna (1992), Herbst and Ablowitz (1992), (1993).

14.8 Reversible systems. Volume-preserving flows

A differential system $\dot{\mathbf{y}} = \mathbf{F}(\mathbf{y})$ is called *reversible* if for some invertible linear map ρ, $\rho(\mathbf{F}(\mathbf{y})) = -\mathbf{F}(\rho(\mathbf{y}))$. The corresponding flow satisfies $\rho\varphi_t = \varphi_t^{-1}\rho$. Reversible systems appear frequently and possess many features in common with Hamiltonian systems (MacKay (1992)). For one-step numerical methods the reversibility condition is $\rho\psi_h = \psi_{-h}\rho$. Stofer (1988a), (1988b) proves that an RK method is reversible if and only if it is symmetric.

The *volume-preserving flow* is another relative of the Hamiltonian flow. The interested reader is referred to MacKay (1992), Scovel (1991), Suris (1987).

References

Abia, L. and Sanz-Serna, J.M. (1993) Partitioned Runge-Kutta methods for separable Hamiltonian problems. *Math. Comput.* **60**, 617–634.

Anosov, D.V. and Arnold, V.I. (eds.) (1988) *Dynamical Systems I*. Springer, Berlin.

Arnold, V.I. (1989) *Mathematical Methods of Classical Mechanics*, 2nd. Edition. Springer, New York.

Beyn, W.-J. (1991) Numerical methods for dynamical systems. In *Advances in Numerical Analysis*, Vol. I, Light, W. ed., Clarendon Press, Oxford, 175–236.

Biesiadecki, J.J. and Skeel, R.D. (1992) Dangers of multiple-time-step methods. Preprint.

Brenan, K.E., Campbell, S.L. and Petzold, L.R. (1989) *Numerical Solution of Initial-Value Problems in Differential-Algebraic Equations*. North-Holland, New York.

Burrage, K. and Butcher, J.C. (1979) Stability criteria for implicit Runge-Kutta methods. *SIAM J. Numer. Anal.* **16**, 46–57.

Butcher, J.C. (1964) Implicit Runge-Kutta processes. *Math. Comput.* **18**, 50–64.

Butcher, J.C. (1987) *The Numerical Analysis of Ordinary Differential Equations*. John Wiley, Chichester.

Calvo, M.P. (1992) Métodos Runge-Kutta-Nyström simplécticos. Tesis Doctoral, Universidad de Valladolid, Valladolid.

Calvo, M.P. and Sanz-Serna, J.M. (1992a) Order conditions for canonical Runge-Kutta-Nyström methods. *BIT* **32**, 131–142.

Calvo, M.P. and Sanz-Serna, J.M. (1992b) Variable steps for symplectic integrators. In *Numerical Analysis 1991*, Griffiths, D.F. and Watson, G.A. eds., Longman, London, 34–48.

Calvo, M.P. and Sanz-Serna, J.M. (1993a) Reasons for a failure. The integration of the two-body problem with a symplectic Runge-Kutta-Nyström code with stepchanging facilities. In *Equadiff 91*, Perelló, C., Simó, C. and Sola-Morales, J. de eds., World Scientific, Singapore.

Calvo, M.P. and Sanz-Serna, J.M. (1993b) The development of variable-step symplectic integrators, with application to the two-body problem. *SIAM J. Sci. Comput.* **14**, 936–952.

Calvo, M.P. and Sanz-Serna, J.M. (1993c) High-order symplectic Runge-Kutta-Nyström methods. *SIAM J. Sci. Comput.* to appear.

Calvo, M.P. and Sanz-Serna, J.M. (1993d) Canonical B-series. *Numer. Math.* to appear.

Candy, J. and Rozmus, W. (1991) A symplectic integration algorithm for separable Hamiltonian functions. *J. Comput. Phys.* **92**, 230–256.

Chan, R.P.K. (1990) On symmetric Runge-Kutta methods of high order. *Computing* **45**, 301–309.

Channell P.J. (1983) Symplectic integration algorithms. *Los Alamos National Laboratory Report*, Report AT-6:ATN 83-9.

Channell, P.J. and Scovel, C. (1990) Symplectic integration of Hamiltonian systems. *Nonlinearity* **3**, 231–259.

Cooper, G.J. (1987) Stability of Runge-Kutta methods for trajectory problems. *IMA J. Numer. Anal.* **7**, 1–13.

Cooper, G.J. and Vignesvaran, R. (1990) A scheme for the implementation of implicit Runge-Kutta methods. *Computing* **45**, 321–332.

Crouzeix, M. (1979) Sur la B-stabilité des méthodes de Runge-Kutta. *Numer. Math.* **32**, 75–82.

Dahlquist, G. (1975) Error analysis for a class of methods for stiff nonlinear initial value problems. In *Numerical Analysis, Dundee 1975*, Watson, G.A. ed., Springer, Berlin, 60–74.

Dekker, K. and Verwer, J.G. (1984) *Stability of Runge-Kutta Methods for Stiff Nonlinear Differential Equations*. North-Holland, Amsterdam.

Dormand, J.R., El-Mikkawy, M.E.A. and Prince, P.J. (1987a) Families of Runge-Kutta-Nyström formulae. *IMA J. Numer. Anal.* **7**, 235–250.

Dormand, J.R., El-Mikkawy, M.E.A. and Prince, P.J. (1987b) High-order embedded Runge-Kutta-Nyström formulae. *IMA J. Numer. Anal.* **7**, 423–430. Corrigendum in **11**, 297.

Dormand, J.R. and Prince, P.J. (1989) Practical Runge-Kutta processes. *SIAM J. Sci. Comput.* **10**, 977–989.

Dragt, A.J. and Finn, J.M. (1976) Lie series and invariant functions for analytic symplectic maps. *J. Math. Phys.* **17**, 2215–2227.

Eirola, T. (1993) Aspects of backward error analysis of numerical ODEs. *J. Comp. Appl. Math.* **45**, 65–73.

Eirola, T. and Sanz-Serna, J.M. (1992) Conservation of integrals and symplectic structure in the integration of differential equations by multistep methods. *Numer. Math.* **61**, 281–290.

Feng, K. (1985) On difference schemes and symplectic geometry. In *Proceedings of the 1984 Beijing Symposium on Differential Geometry and Differential Equations*, Feng, K. ed., Science Press, Beijing, 42–58.

Feng, K. (1986a) Difference schemes for Hamiltonian formalism and symplectic geometry. *J. Comput. Math.* **4**, 279–289.

Feng, K. (1986b) Symplectic geometry and numerical methods in fluid dynamics. In *Tenth International Conference on Numerical Methods in Fluid Dynamics*, Zhuang, F.G. and Zhu, Y.L. eds., Lecture Notes in Physics 264, Springer, Berlin, 1–7.

Feng, K. and Qin, M.Z. (1987) The symplectic methods for the computation of Hamiltonian equations. In *Numerical Methods for Partial Differential Equations*, Zhu, Y.L. and Guo, B.Y. eds., Lecture Notes in Mathematics 1297, Springer, Berlin, 1–37.

Feng, K. and Qin, M.Z. (1991) Hamiltonian algorithms for Hamiltonian systems and a comparative numerical study. *Comput. Phys. Comm.* **65**, 173–187.

Feng, K., Wu, H.M., Qin, M.Z. and Wang D.L. (1989) Construction of canonical difference schemes for Hamiltonian formalism via generating functions. *J. Comput. Math.* **7**, 71–96.

Forest, E. (1992) Sixth-order Lie group integrators. *J. Comput. Phys.* **99**, 209–213.

Forest, E. and Ruth, R.D. (1990) Fourth-order symplectic integration. *Physica D* **43**, 105–117.

Frutos, J. de, Ortega, T. and Sanz-Serna, J.M. (1990) A Hamiltonian, explicit algorithm with spectral accuracy for the 'good' Boussinesq equation. *Comput. Methods Appl. Mech. Engrg.* **80**, 417–423.

Frutos, J. de and Sanz-Serna, J.M. (1992) An easily implementable fourth-order method for the time integration of wave problems. *J. Comput. Phys.* **103**, 160–168.

Gear, C.W. (1992) Invariants and numerical methods for ODEs. *Physica D* **60**, 303–310.

Gladman, B., Duncan, M. and Candy, J. (1991) Symplectic integrators for long-term integration in celestial mechanics. *Celest. Mech.* **52**, 221–240.

Griffiths, D.F. and Sanz-Serna, J.M. (1986) On the scope of the method of modified equations. *SIAM J. Sci. Comput.* **7**, 994–1008.

Guckenheimer, J. and Holmes, Ph. (1983) *Nonlinear Oscillations, Dynamical Systems and Bifurcations of Vector Fields*. Springer, New York.

Hairer, E. (1993) Backward analysis of numerical integrators and symplectic methods. Preprint.

Hairer, E., Iserles, A. and Sanz-Serna, J.M. (1990) Equilibria of Runge-Kutta methods. *Numer. Math.* **58**, 243–254.

Hairer, E., Lubich, Ch. and Roche, M. (1989) *The Numerical Solution of Differential-Algebraic Systems by Runge-Kutta Methods*. Springer, Berlin.

Hairer, E., Murua, A. and Sanz-Serna, J.M. (1993) The non-existence of symplectic multi-derivative Runge-Kutta methods. Preprint.

Hairer, E., Nørsett, S.P. and Wanner, G. (1987) *Solving Ordinary Differential Equations I, Nonstiff Problems*. Springer, Berlin.

Hairer, E. and Wanner, G. (1974) On the Butcher group and general multivalue methods. *Computing* **13**, 1–15.

Hairer, E. and Wanner, G. (1981) Algebraically stable and implementable Runge-Kutta methods of high order. *SIAM J. Numer. Anal.* **18**, 1098–1108.

Hairer, E. and Wanner, G. (1991) *Solving Ordinary Differential Equations II, Stiff and Differential-Algebraic Problems*. Springer, Berlin.

Hénon, M. and Heiles, C. (1964) The applicability of the third integral of motion: some numerical experiments. *Astron. J.* **69**, 73–79.

Herbst, B.M. and Ablowitz, M.J. (1992) Numerical homoclinic instabilities in the sine-Gordon equation. *Quaestiones Mathematicae* **15**, 345–363.

Herbst, B.M. and Ablowitz, M.J. (1993) Numerical chaos, symplectic integrators and exponentially small splitting distances. *J. Comput. Phys.* **105**, 122–132.

Huang, M. (1991) A Hamiltonian approximation to simulate solitary waves of the Korteweg-de Vries equation. *Math. Comput.* **56**, 607–620.

Iserles, A. (1990) Stability and dynamics of numerical methods for ordinary differential equations. *IMA J. Numer. Anal.* **10**, 1–30.

Iserles, A. (1991) Efficient Runge-Kutta methods for Hamiltonian equations. *Bull. Hellenic Math. Soc.* **32**, 3–20.

Iserles, A. and Nørsett, S.P. (1991) *Order Stars*. Chapman & Hall, London.

Iserles, A., Peplow, A.T. and Stuart, A.M. (1991) A unified approach to spurious solutions introduced by time discretization. Part I: Basic theory. *SIAM J. Numer. Anal.* **28**, 1723–1751.

Iserles, A. and Stuart, A.M. (1992) A unified approach to spurious solutions introduced by time discretization. Part II: BDF-like methods. *IMA J. Numer. Anal.* **12**, 487–502.

Jay, L. (1993) Symplectic partitioned Runge-Kutta methods for stiff and constrained Hamiltonian systems. Preprint.

Kirchgraber, U. (1988) An ODE-solver based on the method of averaging. *Numer. Math.* **53**, 621–652.

Lambert, J.D. (1991) *Numerical Methods for Ordinary Differential Equations, The Initial Value Problem*. Wiley, Chichester.

Lasagni, F.M. (1988) Canonical Runge-Kutta methods. *ZAMP* **39**, 952–953.

Lasagni, F.M. (1990) Integration methods for Hamiltonian differential equations. Unpublished manuscript.

Leimkuhler, B. and Reich, S. (1992) Symplectic integration of constrained Hamiltonian systems. Preprint.

Leimkuhler, B. and Skeel, R.D. (1993) Symplectic numerical integrators in constrained Hamiltonian systems. Preprint.

Li, C.W. and Qin, M.Z. (1988) A symplectic difference scheme for the infinite dimensional Hamiltonian system. *J. Comput. Appl. Maths.* **6**, 164–174.

MacKay, R.S. (1992) Some aspects of the dynamics and numerics of

REFERENCES

Hamiltonian systems. In *The Dynamics of Numerics and the Numerics of Dynamics,* Broomhead, D.S. and Iserles, A. eds., Clarendon Press, Oxford, 137–193.

MacKay, R.S. and Meiss J.D. (eds.) (1987) *Hamiltonian Dynamical Systems.* Adam Hilger, Bristol.

McLachlan, R.I. (1992) Symplectic integration of Hamiltonian wave equations. Preprint.

McLachlan, R.I. and Atela P. (1992) The accuracy of symplectic integrators. *Nonlinearity* 5, 541–562.

Maeda, S. (1991) Certain types of Runge-Kutta-type formulas and reproduction of orbits of linear systems. *Electron. Comm. Japan Part III Fund. Electron. Sci.* 74, 98–104.

Marsden, J.E. (1992) *Lectures on Mechanics.* Cambridge University Press, Cambridge.

Menyuk, C.R. (1984) Some properties of the discrete Hamiltonian method. *Physica D* 11, 109–129.

Miesbach, S. and Pesch, H.J. (1992) Symplectic phase flow approximation for the numerical integration of canonical systems. *Numer. Math.* 61, 501–521.

Neishtadt, A.I. (1984) The separation of motions in systems with rapidly rotating phase. *J. Appl. Math. Mech.* 48, 133–139.

Nørsett, S.P. and Wanner, G. (1981) Perturbed collocation and Runge-Kutta methods. *Numer. Math.* 38, 193–208.

Okunbor, D. (1992) Variable step size does not harm second-order integrators for Hamiltonian systems. Preprint.

Okunbor, D. and Skeel, R.D. (1992a) An explicit Runge-Kutta-Nyström method is canonical if and only if its adjoint is explicit. *SIAM J. Numer. Anal.* 29, 521–527.

Okunbor, D. and Skeel, R.D. (1992b) Explicit canonical methods for Hamiltonian systems. *Math. Comput.* 59, 439–455.

Okunbor, D. and Skeel, R.D. (1993) Canonical Runge-Kutta-Nyström methods of orders 5 and 6. *J. Comput. Appl. Math.* to appear.

Olver, P.J. (1986) *Applications of Lie Groups to Differential Equations.* Springer, New York.

Press, W.H., Flannery, B.P., Teukolski, S.A. and Vetterling, W.T. (1989) *Numerical Recipes, The Art of Scientific Computing.* Cambridge University Press, Cambridge.

Pullin, D.I. and Saffman, P.G. (1991) Long-time symplectic integration: the example of four-vortex motion. *Proc. R. Soc. Lond. A* 432, 481–494.

Qin, M.Z. and Zhang, M.Q. (1990) Multi-stage symplectic schemes of two kinds of Hamiltonian systems for wave equations. *Computers Math. Applic.* 19, 51–62.

Qin, M.Z. and Zhu, W.J. (1992) Construction of higher order symplectic schemes by composition. *Computing* 47, 309–321.

Quinlan, G.D. and Tremaine, S. (1990) Symmetric multistep methods for the numerical integration of planetary orbits. *Astron. J.* **100**, 1694–1700.

Reich, S. (1993) Symplectic integration of constrained Hamiltonian systems by Runge-Kutta methods. Preprint.

Ruth, R.D. (1983) A canonical integration technique. *IEEE Trans. Nucl. Sci.* **30**, 2669–2671.

Saito, S., Sugiura, H. and Mitsui, T. (1992a) Butcher's simplifying assumption for symplectic integrators. *BIT* **32**, 345–349.

Saito, S., Sugiura, H. and Mitsui, T. (1992b) Family of symplectic implicit Runge-Kutta formulae. *BIT* **32**, 539–543.

Sanz-Serna, J.M. (1988) Runge-Kutta schemes for Hamiltonian systems. *BIT* **28**, 877–883.

Sanz-Serna, J.M. (1991) Two topics in nonlinear stability. In *Advances in Numerical Analysis*, Vol. I, Light, W. ed., Clarendon Press, Oxford, 147–174.

Sanz-Serna, J.M. (1992a) The numerical integration of Hamiltonian systems. In *Computational Ordinary Differential Equations*, Cash, J.R. and Gladwell, I. eds., Clarendon Press, Oxford, 437–449.

Sanz-Serna, J.M. (1992b) Numerical ordinary differential equations vs. dynamical systems. In *The Dynamics of Numerics and the Numerics of Dynamics*, Broomhead, D.S. and Iserles, A. eds., Clarendon Press, Oxford, 81–106.

Sanz-Serna, J.M. (1992c) Symplectic integrators for Hamiltonian problems: an overview. *Acta Numerica* **1**, 243–286.

Sanz-Serna, J.M. and Abia, L. (1991) Order conditions for canonical Runge-Kutta schemes. *SIAM J. Numer. Anal.* **28**, 1081–1096.

Sanz-Serna, J.M. and Griffiths, D.F. (1991) A new class of results for the algebraic equations of implicit Runge-Kutta processes. *IMA J. Numer. Anal.* **11**, 449–455.

Sanz-Serna, J.M. and Larsson, S. (1993) Shadows, chaos and saddles. *Appl. Numer. Math.* **13**, 181–190.

Sanz-Serna, J.M. and Vadillo, F. (1987) Studies in numerical nonlinear instability III: Augmented Hamiltonian systems. *SIAM J. Appl. Math.* **47**, 92–108.

Sanz-Serna, J.M. and Verwer, J.G. (1989) Stability and convergence at the PDE/stiff ODE interface. *Appl. Numer. Math.* **5**, 117–132.

Scovel, C. (1991) Symplectic numerical integration of Hamiltonian systems. In *The Geometry of Hamiltonian Systems*, Ratiu, T. ed., Springer, New York, 463–496.

Shampine, L.F. and Gladwell, I. (1992) The next generation of Runge-Kutta codes. In *Computational Ordinary Differential Equations*, Cash, J.R. and Gladwell, I. eds., Clarendon Press, Oxford, 145–164.

Siegel, C.L. and Moser, J.K. (1971) *Lectures on Celestial Mechanics*. Springer, Berlin.

Skeel, R.D. (1993) Variable step size destabilizes the Störmer/leap-frog/Verlet method. *BIT* **33**, 172–175.
Skeel, R.D. and Gear, C.W. (1992) Does variable step size ruin a symplectic integrator? *Physica D* **60**, 311–313.
Stofer, D.M. (1988a) Some geometric and numerical methods for perturbed integrable systems. Ph.D. Thesis, Swiss Federal Institute of Technology, Zürich.
Stofer, D.M. (1988b) On reversible and canonical integration methods. Research Report 88-05, Swiss Federal Institute of Technology, Zürich.
Strang, G. (1963) Accurate partial difference methods I: Linear Cauchy problems. *Arch. Rat. Mech. Anal.* **12**, 392–402.
Strang, G. (1968) On the construction and comparison of difference schemes. *SIAM J. Numer. Anal.* **5**, 506–517.
Sun, G. (1992a) Construction of high order symplectic Runge-Kutta methods. Preprint.
Sun, G. (1992b) Symplectic Partitioned Runge-Kutta methods. Preprint.
Suris, Y.B. (1987) Some properties of methods for the numerical integration of systems of the form $\ddot{x} = f(x)$. *U.S.S.R. Comput. Maths. Math. Phys.* **27**, 149–156.
Suris, Y.B. (1988) Preservation of symplectic structure in the numerical solution of Hamiltonian systems. In *Numerical Solution of Differential Equations,* Filippov, S.S. ed., Akad. Nauk. SSSR, Inst. Prikl. Mat., Moscow, 148–160 (in Russian).
Suris, Y.B. (1989) The canonicity of mappings generated by Runge-Kutta type methods when integrating the systems $\ddot{x} = -\partial U/\partial x$. *U.S.S.R. Comput. Maths. Math. Phys.* **29**, 138–144.
Suris, Y.B. (1990) Hamiltonian methods of Runge-Kutta type and their variational interpretation. *Math. Model.* **2**, 78–87 (in Russian).
Suzuki, M. (1991) General theory of fractal path integrals with applications to many-body theories and statistical physics. *J. Math. Phys.* **32**, 400–407.
Varadarajan, V.S. (1974) *Lie Groups, Lie Algebras and their Representations.* Prentice-Hall, Englewood Cliffs.
Wang, D. (1991) Semi-discrete Fourier spectral approximations of infinite dimensional Hamiltonian systems and conservation laws. *Computers Maths. Applic.* **21**, 63–75.
Warming, R.F. and Hyett, B.J. (1974) The modified equation approach to the stability and accuracy analysis of finite difference methods. *J. Comput. Phys.* **14**, 159–179.
Wisdom, J. and Holman, M. (1991) Symplectic maps for the N-body problem. *Astron. J.* **102**, 1528–1538.
Wu, Y.H. (1988) The generating function for the solution of ODE's and its discrete methods. *Computers Math. Applic.* **15**, 1041–1050.
Yoshida, H. (1990) Construction of higher order symplectic integrators.

Phys. Lett. A **150**, 262–268.

Zhong, G. and Marsden, J.E. (1988) Lie-Poisson Hamilton-Jacobi theory and Lie-Poisson integrators. *Phys. Lett. A* **133**, 134–139.

Symbol Index

Roman letters

a_{ij}	RK coefficients *28*
	PRK coefficients for p variables *34*
A_{ij}	PRK coefficients for q variables *34*
b_i	RK weights *28*
	PRK weights for p variables *34*
	RKN weights for v variables *36*
B_i	PRK weights for q variables *34*
c_i	RK abscissae *29*
C_i	PRK abscissae *35*
d	Number of degrees of freedom *1*
D	Dimension of differential system *25*
e	Eccentricity of an orbit *8*
f	Scalar case of **f** *18*
$f(\sigma\nu\rho\tau)$	Elementary differential *50*
f	Right-hand side in partitioned system *34*
	Right-hand side in second-order system *36*
F	Right-hand side in differential system *25*
$\mathcal{F}(\rho\tau)$	Elementary differential *44*
$[\mathcal{F},\mathcal{G}](\beta\rho\tau)$	Elementary differential, partitioned system *48*
g	Scalar case of **g** *18*
g	Right-hand side in partitioned system *34*
h	A constant value of the Hamiltonian *4*
	Step size *25, 27*

SYMBOL INDEX

h_n	Step size (when variable) *25*
$H = H(\mathbf{p}, \mathbf{q}, t)$	Hamiltonian function *1*
I	Time interval *1*
J^{-1}	Matrix in Hamiltonian system *2*
M	Angular momentum *7*
p	Scalar case of $\mathbf{p}$ *3*
$\mathbf{p} = (p_1, \ldots, p_d)$	Momenta in a Hamiltonian system *2*
	Dependent variables in partitioned differential system *34*
$\mathbf{P}_i$	Stage vectors for $\mathbf{p}$ *35*
q	Scalar case of $\mathbf{q}$ *3*
$\mathbf{q} = (q_1, \ldots, q_d)$	Coordinates in a Hamiltonian system *2*
	Dependent variables in partitioned differential system *34*
	Dependent variables in second-order system *36*
$\mathbf{Q}_i$	Stage vectors for $\mathbf{q}$ *35*
r	Radius in polar coordinates *7*
	Order of a numerical method *26*
$\mathcal{R}$	Real line *1*
$\mathcal{R}^{2d}$	$2d$-dimensional Euclidean space *1*
s	Number of stages *28, 34, 36*
S_H	Hamiltonian system with Hamiltonian H *1*
t	Time *1*
t_n	Time level *25*
T	Kinetic energy *3*
$\mathcal{T}$	Period *3*
$\mathbf{v}$	Time derivative in second-order system *36*
V	Potential energy *3*
$\mathbf{y}$	Point in phase space *2*
	Dependent variables in a differential system *25*
$\mathbf{y}^n$	Numerical solution at time level t_n *25*

SYMBOL INDEX

$\mathbf{Y}_i$ Stage vector for **y** *29*

Greek letters

α_{ij} RKN coefficients *36*
$\alpha(\beta\rho\tau)$ Number of monotonic labellings of $\beta\rho\tau$ *48*
$\alpha(\rho\tau)$ Number of monotonic labellings of $\rho\tau$ *44*
$\alpha(\sigma\nu\rho\tau)$ Number of monotonic labellings of $\sigma\nu\rho\tau$ *50*

β_i RKN weights for q variables *36*
$\beta\rho\tau$ Bicolour rooted tree *45*
$\beta\tau$ Bicolour tree *87*

γ_i RKN abscissae *36*
$\gamma(\beta\rho\tau)$ Density of $\beta\rho\tau$ *46*
$\gamma(\rho\tau)$ Density of $\rho\tau$ *41*
$\gamma(\sigma\nu\rho\tau)$ Density of $\sigma\nu\rho\tau$ *50*

θ Polar angle in polar coordinates *7*

ν Frequency *3*

$\rho\tau$ Rooted tree *41*

$\sigma\nu\rho\tau$ Special Nyström rooted tree *48*
$\sigma\nu\tau$ Special Nyström tree *88*

τ Tree *87*

$\phi_{t,\mathbf{F}}$ Flow of autonomous differential system *25*
$\phi_{t,H}$ Flow of autonomous Hamiltonian system *16*
$\Phi_{\mathbf{F}}(t,t_0)$ Solution operator of differential system *25*
$\Phi_H(t,t_0)$ Solution operator of Hamiltonian system *15*
$\Phi(\beta\rho\tau)$ Elementary weight of $\beta\rho\tau$ *46*
$\Phi(\rho\tau)$ Elementary weight of $\rho\tau$ *41*
$\Phi(\sigma\nu\rho\tau)$ Elementary weight of $\sigma\nu\rho\tau$ *50*

ψ Transformation *19*
ψ' Jacobian matrix of transformation ψ *20*
$\psi_{h_n,\mathbf{F}}$ Numerical method, autonomous case *26*

$\psi_{h_n,H}$	Numerical method, autonomous Hamiltonian case *26*
$\Psi_{\mathbf{F}}(t_{n+1}, t_n)$	Numerical method *25*
$\Psi_H(t_{n+1}, t_n)$	Numerical method, Hamiltonian case *26*
ω	Angular frequency *3*
Ω	Phase space *1*, *25*
$\Omega \times I$	Extended phase space *1*, *25*

Binary operations

$\wedge$	Exterior product *21*
$\otimes$	Kronecker (tensor) product *61*
$\{\cdot, \cdot\}$	Poisson bracket *155*
$[\cdot, \cdot]$	Commutator *157*

Operators

∇	Gradient *2*
$\dot{}$	Time derivative, e.g. $\dot{H}$ *3*

Index

Abscissae
  Of PRK methods *35*
  Of RK methods *29*
Absolute stability *27*
Adjoint method
  Definition *38*
  Of a diagonally implicit, symplectic RK method *101*
  Of an explicit, symplectic PRK method *105*
  Of RK, PRK and RKN methods *39*
Adjoint representation *157*
Algebraic equations in implicit methods
  Existence of solutions *33*
  Practical solution of *61–65*
Algebraic stability *75*
Aliasing *10*
Angular frequency *3*
  See also Frequency, Period
Angular momentum *7*
  See also Conservation of angular momentum
Apocentre of an orbit *7*
Area-preserving transformation
  Checking preservation of area *19–21*
  Definition *16*
Artificial satellite *8*
A-stability *28, 100*
Autonomous system *2, 16, 26, 43, 46*

Backward error analysis *129–136, 140*
Backward Euler rule *33, 38, 70, 139*
Baker-Campbell-Hausdorff (BCH) formula *160–164, 172*
BCH formula *see* Baker-Campbell-Hausdorff formula
Bicolour rooted tree *45*
Bicolour tree *87*
Black vertex *46*
Branches of solutions *33*
B-series *180*
B-stability *28, 100, 102*

Canonical *see* Symplectic
Canonical theory of the order *136, 150–153*
Central force *6*
Centre *18, 70, 129, 141*
Chaotic solution *14*
Collocation methods
  Definition *30*
  Using the collocation polynomial to start the iteration *63*
Commutator of operators *157*
Composition of methods
  Definition *37*
  Of leap-frog methods *109–110, 165–170, 172*
  Of order 1, explicit, symplectic PRK methods *105–107*

Of order 2, explicit, symplectic RKN methods *172*
Of order 3, explicit, symplectic PRK methods *108*
Of the midpoint rule *100–102, 165–170*
Yoshida's approach *165–170*
See also Fractional-step method
Concatenation of methods see Composition of methods
Configuration space *6*
Conjugate symplectic method *181*
Conservation of angular momentum *7–8, 56, 136*
Conservation of area see Area-preserving transformation
Conservation of energy *2–13, 136–140*
Conservation of invariant quantities *136–140*
See also Conservation of energy, Conservation of angular momentum
Conservation of volume *23, 188*
Consistency *26*
Constrained system *182–184*
Coordinates *2*

Degrees of freedom *1*
Density
Of a bicolour rooted tree *46*
Of a rooted tree *42*
Of a special Nyström rooted tree *50*
Diagonally implicit method
Of PRK type *35*
Of RK type *29*
Symplectic *100*
Symplectic, high-order *165–170*
Differential-algebraic equations (DAEs) *139, 183*
Differential form *20*
Differential operator *156–157*
Differential system *25*

See also Autonomous system, Differential-algebraic equations, Hamiltonian system, Partial differential equations, Reversible system, Stiff system
Double harmonic oscillator *5*
Dynamics *18, 70, 138, 141*
See also Chaotic solution Equilibrium, Ergodic trajectory, Libration, Limit cycle, Periodic solution, Poincaré recurrence, Poincaré section, Precession of the pericentre, Quasiperiodic solution, Rotation, Separatrix, Stochastic solution, Torus

Eccentricity of an orbit *8*
Elementary differential
For a bicolour rooted tree *48*
For a rooted tree *44*
For a special Nyström rooted tree *51*
Independence in Hamiltonian problems *81*
Elementary Hamiltonian *153*
Elementary weight
For a bicolour rooted tree *46*
For a rooted tree *41*
For a special Nyström rooted tree *50*
Of the stages *82*
Embedded pair *54*
End-vertex *50*
Energy *2*
Equilibrium *4, 18*
See also Centre, Saddle, Sink, Source, Spiral point
Equivalent stages *82*
Ergodic trajectory *6*
Error constants *112–113, 116, 174*
Euler rule *25, 27, 38, 70, 75, 131*

INDEX

See also Backward Euler rule
Exact symplectic transformation *149*
Explicit methods
  Not suitable for stiff problems *28*
  Of PRK type *35*
  Of RK type *29*
  Of RKN type *36*
Explicit midpoint rule *186*
Extended phase space *1*
Exterior product *21-22*

Fast Fourier transform *10*
Fat vertex *48*
Father (in a rooted tree) *41*
Flow
  General *25*
  Hamiltonian *16*
Fluid mechanics *100*
Fourier analysis see Fast Fourier transform
Fractional-step method *161-170*
Frequency *3*
  See also Angular frequency, Nyquist frequency, Period
FSAL (First same as last) *56*
Functional iteration *62-68*, *115*, *173*

Gauss methods
  Are symmetric *39*
  Are symplectic *99*
  Alternative writing *61*
  Definition *31-33*
  Induce RKN methods *37*
  Numerical experiments *65-68*, *115-124*, *140*, *173-177*
Generating function *136*, *143-153*
Generic properties *18*
Global error *27*
Gradient *2*

Hamilton-Jacobi equation *146-148*
Hamiltonian system (general) *184-185*
Hamiltonian system (standard) *1*
  See also Locally Hamiltonian system, Separable Hamiltonian
Harmonic oscillator *3*, *17*, *70*, *132*, *138*, *144-146*, *159*
  See also Double harmonic oscillator
Hénon-Heiles Hamiltonian *12-14*, *121-123*, *176*
Homogeneous form of the order conditions for symplectic methods *95-98*

Implementation
  Main reference *53-68*
  Of diagonally implicit, symplectic RK methods *101*
  Of explicit, symplectic PRK methods *103-105*
  Of explicit, symplectic RKN methods *111*
  Of the midpoint rule *31*
Implicit methods
  Definition *26*
  For stiff systems *28*
  Implementation *61-65*
  Of RK type *29*
Initial guess for iteration *63*
Integrable Hamiltonian *13*, *141*, *161*

Jacobi condition *155*, *185*
Jacobian *19-20*, *22-23*, *62*, *67-68*

KAM theory *14*, *141*
Kepler's problem *6-8*, *56-60*, *117-121*, *125-126*, *174-176*
  See also Modified Kepler problem
Kinetic energy *3*, *76*

Relativistic 77
Korteweg-de Vries equation *187*
Kronecker product of matrices *61*
Kuntzmann-Butcher methods *31*

Leap-frog method *107, 165*
Leibniz rule *155, 184*
Libration *4*
Lie algebra *155–157*
Lie bracket *164*
Lie operator *156*
Lie series *157–159*
Limit cycle *18*
Liouville theorem *17, 19, 23*
Lipschitz constant *28, 33, 62, 64, 83*
Lobatto methods *103, 179*
Local error
   Definition *26*
   Of PRK methods *46–48*
   Of RK methods *43–45*
   Of RKN methods *50–52*
Local error estimation *53–54*
Local extrapolation *54*
Locally Hamiltonian system *19, 22*

Mass matrix *77*
Meagre vertex *48*
Midpoint rule (implicit) *31, 71, 75, 80, 101, 148, 165, 181–182*
   See also Explicit midpoint rule
Modified equations, method of *129*
Modified Hamiltonian *129–136, 161–163, 165–170*
Modified Kepler problem *8–12, 76, 123–124, 176*
Momenta
   In polar coordinates *7*
   Of a Hamiltonian system *2*
Monotonic labelling *44, 48, 50*
Multiderivative numerical method *149, 180*

Multistep numerical method *185–186*

Newton iteration *62–68, 100*
Numerical experiments
   Backward error interpretation *129–131*
   Conservation of energy *140*
   Constant vs. variable step sizes *55–60*
   Gauss order-4 *65–68*
   High-order methods *173–177*
   Symplectic vs. nonsymplectic methods *69–71, 115–124, 173–177*
   Variable steps for symplectic integration *124–127*
Numerical method
   Definition *25*
   See also Error constants, Fractional-step method, Implementation, Multiderivative numerical method, Multistep numerical method, Optimization of a numerical method, Order of a method, Partitioned Runge-Kutta method, Runge-Kutta method, Runge-Kutta-Nyström method, Symmetric method, Symplectic method
Nyquist frequency *10*

Optimization of a numerical method *55, 112, 115–116, 171, 173–174*
Order conditions
   Are independent *87*
   For PRK methods *45–46*
   For RK methods *41–43*
   For RKN methods *48–50*
   For symplectic PRK methods *91–93, 95–96*

For symplectic RK methods
  88-91, 97
For symplectic RKN methods
  93-95, 97
See also Canonical theory of
  the order, Simplifying
  assumption
Order of a method
  Definition 26
  Order of the adjoint method 38
  Symmetric methods have even
    order 38
  When applied to Hamiltonian
    problems 82
  See also Order conditions
Order of a rooted tree 41

Partial differential equations 28,
  162, 186-187
  See also Hamilton-Jacobi
    equation
Partitioned Runge-Kutta (PRK)
  method
  Adjoint 39
  Definitions 34
  Local error 46-48
  Order conditions 45-46
  See also Implementation,
    Numerical experiments,
    Symplectic PRK method
Partitioned system 34
Pendulum 3-5, 130, 139, 182-184
Pericentre of an orbit 7
Period 3, 5, 7, 8
Periodic solution 3-12, 18
Periodogram 10, 124
Phase space 1
Planetary motion 164, 186
  See also Kepler's problem,
    modified Kepler problem
Poincaré generating function
  145-146, 148
Poincaré integral invariant 22
Poincaré recurrence 23
Poincaré section

Computation 123
Definition 10
Examples 10, 13, 122
Poisson bracket of functions
  (general) 184
Poisson bracket of functions
  (standard) 155
Poisson bracket of vector fields
  164
Poisson integrator 185
Poisson map 185
Potential energy 3, 76
Precession of the pericentre 9
PRK method see Partitioned
  Runge-Kutta method

Quadrature rule 30, 99, 102
Quasiperiodic solution 6, 9-10, 14

Radau methods 103
Redundant stages 80, 82-85, 100,
  103, 105, 106, 111
Relativity 77
Reversible system 188
RK method see Runge-Kutta
  method
RKN method see
  Runge-Kutta-Nyström method
Root of a rooted tree 41
Rooted tree 41
Rotation 4
Runge-Kutta (RK) method
  Adjoint 39
  Definitions 28
  Local error 43-45
  Order conditions 41-43
  With higher derivatives 149,
    180
  See also Backward Euler rule,
    Euler rule, Gauss methods,
    Implementation, Lobatto
    methods, Midpoint rule,
    Numerical experiments,
    Radau methods, Symplectic

RK method, Trapezoidal
  rule
Runge-Kutta-Nyström (RKN)
  method
  Adjoint *39*
  Definitions *36*
  Induced by a PRK method *37,
    52*
  Local error *50–52*
  Order conditions *48–50*
  See also Implementation,
    Numerical experiments,
    Störmer-Verlet method,
    Symplectic RKN method

Saddle *18*
Second-order system *36, 77*
Semigroup property *16*
Separable Hamiltonian *76*
Separatrix *4*
Simplifying assumption *49, 80,
    170*
Sink *18*
Solution operator
  General *25*
  Hamiltonian *15*
Son (in a rooted tree) *41*
Source *18*
Special Nyström rooted tree *48*
Special Nyström tree *88*
Spiral point *70, 131*
Splitting method see
  Fractional-step method
Spurious solutions
  Main reference *33*
  May be symplectic *74*
Stability function *75*
Stability of a numerical method
  see Absolute stability,
  Algebraic stability, A-stability,
  B-stability, Stability function
  Stiff system
Stages
  Elementary weight *82*
  In PRK methods *35*
  In RK methods *28*
  In RKN methods *36*
  See also Equivalent stages,
    Redundant stages
Step *25*
Step size *25*
  See also Variable step sizes
Stiff system *27, 29, 34, 62, 100,
    162*
Stochastic solution *14*
Stokes theorem *143*
Stopping criterion when solving
  algebraic equations *64*
Störmer-Verlet method *112, 165*
Structure matrix *185*
Superfluous tree *88*
Symmetric method *38, 75,
    101–102, 106, 108, 148, 163,
    165–171, 186*
Symplectic transformation *16, 21*
  See also Area-preserving
    transformation, Exact
    symplectic transformation,
    Generating function, Poisson
    bracket
Symplectic method
  Based on generating functions
    147
  Cannot exactly conserve energy
    138
  Definition *69*
  Properties *129–141*
  See also Conjugate symplectic
    method, Generating
    function, Poisson integrator,
    Symplectic PRK method,
    Symplectic RK method,
    Symplectic RKN method
Symplectic PRK method
  Backward error interpretation
    129–136
  Canonical theory of the order
    150–153
  Conservation properties
    136–140

INDEX

Definition *76*
For nonseparable Hamiltonians *179*
Generating function *150*
High-order *165–170*
Order conditions *91–93, 95–96*
Specific explicit methods *103–110, 165–170*
Symplectic RK method
Backward error interpretation *129–136*
Canonical theory of the order *150–153*
Conservation properties *136–140*
Definition *72*
Generating function *149*
High-order *165–170*
Order conditions *88–91, 97*
Specific methods *99–103, 165–170*
Symplectic RKN method
Backward error interpretation *129–136*
Canonical theory of the order *150–153*
Conservation properties *136–140*
Definition *77*
Generating function *150*
High-order *165–172*
Order conditions *93–95, 97*
Specific methods *110–113, 165–172*

Tableau
Of a PRK method *34*
Of an RK method *28*
Of an RKN method *36*
Time level *25*
Time scale *27*
Tolerance (algebraic equations) *64*
Tolerance (local error) *53*
Torus *5, 9, 12, 14*
Trapezoidal rule *181–182*

Tree *87*

Variable step sizes *53–54, 56–60, 115–127, 133–134, 173–177*
Vertex of a graph *41*
Volume-preserving flow *188*

Wave equation *187*
Wedge product *see* Exterior product
Weight
Is $\neq 0$ for symplectic RK methods *100*
Of a quadrature rule *30*
Of an RK method *29*
Of an RKN method *37*
White vertex *46*
W-transformation *102*